METHODS FOR RESEARCH ON THE ECOLOGY OF SOIL-BORNE PLANT PATHOGENS

by

Leander F. Johnson
University of Tennessee
Knoxville, Tennessee

and

Elroy A. Curl
Auburn University
Auburn, Alabama

Burgess Publishing Company
426 South Sixth Street • Minneapolis, Minnesota 55415

QR
111
.J62
1972

Copyright © 1972 by Burgess Publishing Company

All rights reserved. No part of this book may be reproduced in any form whatsoever, by photograph or mimeograph or by any other means, by broadcast or transmission, by translation into any kind of language, nor by recording electronically or otherwise, without permission in writing from the publisher, except by a reviewer, who may quote brief passages in critical articles and reviews.

Printed in the United States of America
Library of Congress Catalog Card Number 77-176196
SBN 8087-1016-8

1   2   3   4   5   6   7   8   9   0

CONTENTS

Chapter

Introduction	vii
1. Collection of Soil Samples	1
A. Variation and Distribution of Microorganisms in Soil	1
B. Depth and Method of Sampling	3
C. Storage of Soil Samples	3
D. Sampling Procedure	4
2. Isolation of Groups of Microorganisms from Soil	6
A. Isolation of Bacteria from Soil	7
B. Isolation of Actinomycetes from Soil	12
C. Isolation of Fungi from Soil	14
D. Techniques for Isolating Specific Groups of Fungi from Soil	26
E. Isolation of Algae from Soil	33
3. Enumeration and Observation of Microorganisms *in situ*	36
A. Direct Microscopic Examination of Soil	36
B. Buried Slides	39
C. Buried Nylon Gauze	41
D. The Impression Slide	41
E. Counting Soil Algae	42
F. Immunofluorescent Staining Technique	43
G. Flotation of Spores	43
H. Hydrofluoric Acid Extraction	43
I. Soil Sectioning	44
4. Microorganisms in the Rhizosphere	48
A. Isolation from the Rhizosphere	49
B. Isolation from the Rhizoplane	51
C. Special Devices for Collecting Rhizosphere Samples	52
D. Study of Organisms in the Rhizosphere without Isolation in Pure Culture	55
5. Isolation of Plant Pathogens from Soil	58
A. Bacteria	59
B. *Cylindrocladium scoparium*	60

- C. *Fomes annosus* 60
- D. *Fusarium* 60
- E. *Geotrichum candidum* 62
- F. *Helminthosporium sativum* 62
- G. *Phymatotrichum omnivorum* 62
- H. *Phytophthora* 63
- I. *Pythium* 65
- J. *Rhizoctonia solani* 67
- K. *Sclerotium* 68
- L. *Streptomyces scabies* 69
- M. *Thielaviopsis basicola* 70
- N. *Verticillium albo-atrum* and *V. dahliae* 71

6. Isolation of Plant Pathogens from Roots 73
   - A. Specific Plant Pathogens 74
   - B. Wood Rotting Fungi 78
   - C. Mycorrhizal Fungi 79
   - D. Viruses 80

7. Control of Soil Environment 82
   - A. Soil Moisture 82
   - B. Soil Temperature 85
   - C. Soil Atmosphere 85
   - D. Soil Fertility and pH 89
   - E. Soil Sterilization 89

8. Extraction of Soil Solutions 92
   - A. Water Extracts of Soil 92
   - B. Vacuum Filtration 96
   - C. Centrifugation 97
   - D. Displacement with Alcohol 99
   - E. Pressure-Membrane Apparatus 99

9. Respiration and Enzyme Activity 102
   - A. Respiration 102
   - B. Enzyme Activity 108

10. Growth and Survival 116
    - A. Assessing Growth in Soil 116
    - B. Survival 118
    - C. Soil Fungistasis 121
    - D. Competitive Saprophytic Ability (CSA) 124
    - E. Inoculum Density and Potential 125

11. Root Exudates 130
    - A. Collection 130
    - B. Detection and Analysis of Specific Substances 135
    - C. Effect on Microorganisms 135

12. Screening Soil Microorganisms for Antagonism 139
    A. The Bacterial Agar Plate 140
    B. Layers of Agar 140
    C. Spray Techniques 142
    D. Soil Enrichment 142
    E. Predation and Parasitism Among Microorganisms 144
    F. Bacteriophages 148
    G. Actinophages 150
13. Testing for Antibiotic Activity 153
    A. Agar Culture 153
    B. Assay of Culture Filtrates or Antibiotics 157
14. Production, Stability, and Activity of Antibiotics in Soil 167
    A. Recovery of Antibiotics Added to Soil 168
    B. Isolation of Antibiotics Produced in Soil 169
    C. Direct Bioassay of Soil for Presence of Antibiotics 171
    D. Special Techniques for Demonstrating Antibiotic
       Production in Soil 173
15. Biological Control 176
    A. Use of Pure Cultures of Antagonistic Organisms 177
    B. Organic Amendments 180
    C. Crop Rotation 183
    D. Sterilization and "Pasteurization" of Soil, and Stimulation
       of Antagonists by Chemicals or Cultural Practices 184
16. Culture Media 187
    A. General Purpose Media 187
    B. Media for Isolating Bacteria and Actinomycetes from Soil . 191
    C. Media for Isolating Actinomycetes from Soil 193
    D. Media for Isolating Fungi from Soil 195
    E. Selective Media for Isolating Specific Groups or
       Genera of Microorganisms from Soil or Plants 197
    F. Fermentation Media for Antibiotic Production 203
    G. Culture Solutions for Higher Plants 206

Bibliography 209
Index—Subject 237
Index—Senior Author 243

INTRODUCTION

To encourage and aid research on basic aspects of the ecology of soil organisms, a compilation of techniques was made in 1959 by the present authors along with J. H. Bond and H. A. Fribourg under the title, "Methods for Studying Soil Microflora – Plant Disease Relationships." This manual, though limited in scope, brought to the attention of the plant pathologist certain techniques hitherto used almost exclusively by soil microbial ecologists, and bridged the gap to some extent between the sciences of plant pathology and soil microbiology.

Recently, increased emphasis in research on soil-borne diseases and biological control has led to the development of new and better methods and has made obsolete many of the older techniques. Some of the better methods outlined in "Methods for Studying Soil Microflora – Plant Disease Relationships" are retained in the present volume. Included also are new and improved techniques and chapters which provide methods useful to plant pathologists. Details of old, seldom-used methods are not outlined, but references to many of them are cited. Pertinent data to illustrate specific points along with illustrations and diagrams are added for clarity.

This compilation is designed primarily to be used as a laboratory manual by professional scientists and students engaged in research on soil-borne plant diseases. It can be used also as a reference manual for specialized courses in plant pathology, soil microbiology, and microbial ecology.

Many methods are outlined in detail, so that it is not necessary to refer to original references. In some cases, only a generalized procedure is given where a detailed outline might require considerable background information, or where detailed procedures were not included by the author in his original publication. For further information the reader is referred to the original publication and others of a similar nature. The method of presentation is varied according to the subject under consideration. Procedures for many of the methods are outlined in sequential order, so that they can be followed step by step. Others are difficult to outline because of the variability encountered or because of the nature of the research involved. Where the procedures are not outlined in sequential order, a generalized historical review is given, covering procedures used and the results obtained.

Many of the methods described in this manual have not been tested by us and no attempt is made to influence the reader in his selection. We occasionally indicate our preference where two or more methods are presented for accomplishing a specific purpose, especially if we have had personal experience with the techniques involved. Some methods may be recommended if they possess a distinct advantage over others as determined by experimental data found in the literature. Often, the reader can select a method easily by considering the time and equipment at his disposal. Many techniques are compiled elsewhere and no attempt is made to include all methods that might be useful.

A special note of thanks is extended to Dr. Coyt Wilson, Director of the Virginia Agricultural Experiment Station, whose encouragement and personal assistance made possible the formation of Regional Project S-26, and later the publication of this manual. We are grateful for permission to reproduce published illustrations by various authors, copyright owners, and publishers. Thanks are due Mrs. Grace Early for aid in making many of the line drawings and photographs and to Miss Barbara Jones for invaluable aid in library work and preparation of the manuscript. We are especially indebted to Drs. L. S. Bird, R. Rodriguez-Kabana, and G. C. Papavizas for critical readings of the manuscript. Finally, we wish to thank the technical committee members of Regional Project S-26 for encouragement and advice during the preparation of the manuscript.

L. F. J.
E. A. C.

CHAPTER 1
COLLECTION OF SOIL SAMPLES

When quantitative studies of soil microorganisms are made, several problems in soil sampling may arise. In order to solve them the worker must be aware of the natural variation within experimental plots. He must make a decision about the number of samples to collect, the sites of collection, depth of sampling, size of each sample, and the physical method of collection. He must also decide whether to composite and mix the samples after collection and whether storage of samples before analysis might affect his results. Decisions finally reached are based on the nature of the area under study, the method of microbial analysis used, and the facilities available. These factors vary considerably from place to place with the result that "standardized methods" for collecting soil samples are impractical.

A. VARIATION AND DISTRIBUTION OF MICROORGANISMS IN SOIL

Early workers generally assumed that two soils of the same type, subjected to like treatments in the same locality, contain at a given depth approximately the same number and kind of microorganisms. It has since been shown that microbial variation is considerable, even between samples taken a few inches apart. The nature of the distribution of individuals and groups or organisms in the soil account for this variation. The microflora are rarely free in the liquid phase of the soil but occur largely as colonies attached to clay, humus, and organic matter particles. An organic particle (for example, a small rootlet) may have associated with it a number of organisms distinctly different from those associated with an adjacent particle. The microbial population of a soil should be considered as a composite of these "microecological environments."

Theoretically, a perfect Poisson distribution curve of the frequency of isolation of a particular soil organism would be obtained if the organism were distributed completely at random and if there were no influence of one organism on another. Thus, the presence of one individual in a sampling unit would not influence the probability of isolating others. Although the occurrence of individuals distributed completely at random is rarely encountered in nature, the Poisson distribution makes a useful yardstick with which to compare actual distributions. Thus, Nash and Snyder (1962) found that the number of colonies of *Fusarium* isolated on individual plates containing a selective medium followed a Poisson distribution, which was evidence for the accuracy of their plate-count technique. For details of the application of statistical techniques for investigating spatial distributions, and for analyzing counts from field experiments, see Healy (1962).

To determine the minimum number of samples necessary to obtain adequate data, a preliminary study of the microbial variation or uniformity should be made. In general, the larger the number of random samples taken from an area, the more representative of the area sampled will be the composite of these samples. Compositing the samples may yield results representative of the total area under study, but variation within the area will not be determined. Analysis of individual core samples taken at random should indicate the extent of this variation and statistical treatment of such data can be made. Fungal counts (as measured by direct plate counts) from soil samples, taken from a presumably homogeneous experimental area, frequently show variations of the order of 1:13 (Rose and Miller, 1954). The main source of error is the variation between cores in the field. Much less is due to subsampling and plating methods in the laboratory.

A study of variation in number and kinds of fungi within cultivated field plots was made by Schmitthenner and Williams (1958). Previously, two soil samples per plot were analyzed, but it was found that in many instances the differences between the paired samples were as great or greater than differences between treatments. By increasing the number of samples taken from each plot the efficiency of isolation (determined by a decrease in variance of sample means) could be increased considerably. Pooling the samples further increased the efficiency and decreased the time and equipment needed for analysis. For isolating fungi, 8-32 cores of soil were collected to a depth of 10 cm with a soil sampler. These cores were pooled and 10 dilution plates were prepared from each of the pooled samples.

Core samples obtained from one of the field experimental units constitute samples of this experimental unit, and not replications. To have replication of the soil samples, cores must be obtained from two or more of the plot replications of each of the field treatments under consideration. In order to relate the effect of field treatments (for example, fertilization, pH changes, organic amendments, cropping sequences, etc.) to numbers and kinds of microorganisms present, replications must be maintained to furnish information relative to field variation for the treatment in question. Thus, a composite soil sample may be taken from field replication 1 of a certain treatment and another composite sample is taken from field replication 2 of the same treatment. If these two composite samples are kept separate and analyzed separately, the two replications of the field treatment are maintained and valid statistical measurements comparing treatment means may be made.

B. DEPTH AND METHOD OF SAMPLING

Microorganisms vary considerably in number and kind with depth of soil. In general, fewer organisms are found with increasing depth, but in some virgin soils numbers of bacteria and actinomycetes increase with depth to the upper limits of the B horizon, then decrease with further depth. Since soils vary in the nature and extent of the different stratified layers, it is important that the samples be referred to both definite depths and definite horizons.

In undisturbed soils the most commonly used method of collecting samples consists of digging pits in the area to be sampled, and collecting the samples with trowels or sterile vials pushed horizontally into the prepared face at various depths. Samples should be taken from the bottom of the pit upwards to avoid contamination of the soil face by material falling from the upper part of the profile. According to the nature of the information desired, samples at specific depths may be composited before analysis. Some investigators utilize soil augers pressed or drilled vertically into the soil to obtain samples. The soil core thus obtained is separated into component fractions for microbial analysis. Lipman and Martin (1918) presented experimental evidence that the ordinary removal of samples vertically from soil by augers seems to introduce no contamination from one depth of soil to another, as the auger passes downward. The soil microflora are apparently so large, so characteristic, and so firmly established and adapted to the conditions under which they are found that the introduction of relatively small numbers of contaminating organisms into the sample is without perceptible effect on the isolation of the characteristic microflora at specified depths. A number of probes or "soil-sampling tubes" designed especially for soil microbiological investigations have been described (Unger, 1957; Gilmore, 1959; Rozhdestvenskii, 1961).

Vertical samples are often collected from cultivated soils. Usually, cores of soil are taken to a specified depth (10-20 cm), and the soil within the core is mixed. Cores from the same field replication may be composited before analysis.

C. STORAGE OF SOIL SAMPLES

Golebiowska (1957) in studies on the effect of storage under moist and dry conditions in the laboratory found that numbers obtained by the dilution plate method decreased considerably up to the 14th day in both moist and dry conditions. After this time, an "equilibrium" was reached; that is, total numbers isolated did not vary appreciably during some 7 months of the experiment. Slightly fewer colonies were obtained from the dry soil than from the moist soil during the period of equilibrium. Stotzky et al. (1962) reported changes in the populations in soil samples stored in polyethylene bags for 3 months with little moisture loss during this period. During storage, numbers of bacteria and fungi decreased considerably, but numbers of actinomycetes increased. If soil is left to dry at room temperature, both numbers and kinds of fungi begin to decrease within 24 hr (Manka, 1964). Microbial counts of soil samples stored at -15 C by Grossbard and Hall (1964) differed significantly from counts of unfrozen samples.

The microflora obtained after storage under constant conditions could represent a static biological system. Variability in the results of analyses of samples collected subsequently from the same area might be decreased. Yet, it is likely that during storage many species of the active soil microflora might be decreased to a point where their isolation is not effected. Furthermore, storage in the laboratory, whether in a moist or dry condition, is extremely artificial; such environmental conditions are almost never encountered in nature. It is recommended that analysis be made immediately after collection of samples; thus the results will represent the actual microflora existing at the time of sampling.

D. SAMPLING PROCEDURE

The following procedure may be used for taking soil samples for general quantitative and qualitative determinations of microorganisms in replicated field plot experiments.

1. A metal soil-sample tube with appropriate graduations is used. If sampling at a considerable depth is desired, a soil auger may be used. The instrument is washed thoroughly before starting the sampling procedure.
2. In cultivated soils, samples are taken usually to a depth of 15 cm and the soil is shaken directly into clean plastic bags or other suitable containers.
3. At least five samples are taken at random from each replication of each treatment. It should be recognized that the larger the number of random cores taken, the more representative of the total soil mass will be the composite of the samples. More than five samples should be taken from large plots and from plots within which visible differences in soil color, texture, type or cropping history exist.
4. The samples are kept in a cool place during transportation to the laboratory. Procedures for microbial analysis are begun as soon as possible after sampling, preferably the same day.
5. The five or more samples from each replication are brought together into one composite sample which is mixed thoroughly and divided several times. Finally, a weighed aliquot of each composite sample is taken for analysis.

The procedure given above is applicable to the dilution-plate method of microbial analysis. Some modifications of this procedure may be necessary if other methods are used to count and/or isolate soil microorganisms.

CHAPTER 2
ISOLATION OF GROUPS OF MICROORGANISMS FROM SOIL

Methods described in this chapter are those that yield a relatively broad spectrum of bacteria, actinomycetes, fungi, or algae in pure culture. All of the methods described have one limitation in common, i.e., only members of the particular group of soil organisms that will grow on the medium chosen are obtained. Because of this limitation, no single method has been devised that will yield a true random selection of the total population of the living soil microflora. Results from direct microscopic counts have shown that only 10 to 20% of the total bacterial soil population is obtained by techniques such as "soil dilution." It is assumed, therefore, that many bacteria and some fungi, because of their diverse nutritional requirements, are not isolated from soil in pure culture.

It is difficult to estimate accurately the number of fungi in soil by dilution techniques since species that sporulate abundantly are most often isolated and mycelia of some fungi tend to break up into viable fragments more readily than others. Thus, the soil-dilution technique may produce misleading information about the distribution of species and about the total number of viable propagules of fungi per gram of soil. In attempts to arrive at a more satisfactory estimate of fungal populations, several techniques have been devised which are based on isolating actively growing hyphae from the soil, but even these sometimes yield misleading information because fungi that produce "fast-growing" hyphae, and hyphae capable of growing through an air space onto the agar medium, are most often isolated.

While these methods might leave one uncertain as to the absolute numbers of organisms in soil, they are useful for general purposes and necessary for supplying the research worker with a number of cultures of different kinds of organisms. In most cases results from replicated soil samples tend to agree within random sampling expectation. Therefore, it is possible to compare the results obtained by the same technique from a number of soil samples taken at different locations or from samples previously treated in different ways.

A. ISOLATION OF BACTERIA FROM SOIL

Unless specific kinds of bacteria are desired, such as nitrogen-fixing, anaerobic, or distinct genera that can be isolated by enrichment methods only, dilution of soil with water and subsequent culturing of portions of the soil suspension on agar media is the basic technique for obtaining bacteria from soil. Details of two methods of handling soil dilutions are given below with a section on culture media used.

1. Dilution-plate technique:

Bacterial colonies obtained on soil-dilution plates can be counted, and an estimate of the total population in the soil sample assayed can be calculated. This estimate refers to the number of viable cells in the sample capable of growing on the agar medium employed in the test. Details of the procedure can be varied according to the preference of the individual worker. The following procedure may serve as a guide:

Soil to be diluted is sifted through a sieve with 2-mm pores. Three samples (5-10 g each) are weighed in previously weighed metal or glass containers and dried overnight in an oven at 105-110 C. The dried samples are then reweighed and the moisture content of the soil sample calculated.

A 25-g sample of the sieved soil (determined on a dry-soil basis) is placed in a graduated cylinder. Water is added so that a total volume of 250 ml is reached. The suspension is stirred and poured into a 1000-ml Erlenmeyer flask. The flask containing the suspension is shaken on a mechanical shaker for 30 min. Ten-ml of this suspension are immediately drawn (while in motion) into a sterile 10-ml pipette and transferred into a 90-ml sterile water blank. Screw-cap medicine bottles (200 ml) with flat-oval shape are preferred as the water blank containers. Ten-ml samples are then transferred immediately through successive 90-ml sterile water blanks until the desired final dilution is reached. Each suspension is shaken by hand for a few sec, and is in motion while being drawn into the pipette. The sample should not be allowed to remain in any dilution for more than 10 min.

The preceding method for making soil dilutions yields dilutions of soil in water of 1:10, 1:100, 1:1000, 1:10,000, 1:100,000, 1:1,000,000, 1:10,000,000, etc. In an intermediate dilution is desired, it may be obtained by varying the volume of the water blank and the amount transferred. For example if a dilution of 1:20,000 is desired, 5 ml of the 1:1000 dilution is transferred into a 95-ml water blank. Proper dilution will vary with the soil used and should be determined by preliminary trials. For many cultivated soils, a dilution of 1:1,000,000 is suitable.

One-ml of the desired dilution is transferred aseptically into each of several petri dishes, and 12-15 ml of an appropriate agar medium, cooled to just above the solidifying temperature, is added to each dish. The dishes are rotated by hand in a broad, swirling motion so that the diluted soil is dispersed in the agar medium.

If surface colonies only are desired, 0.5 ml of the final dilution is transferred to the surface of hardened agar medium in each petri dish. The suspension can be spread over the agar surface by rotation or with a glass rod. If this method is used, it is necessary to prepare the agar plates 2-4 days previously, to ensure that the agar surface is dry and that all the liquid will be readily absorbed. Failure to use partly dried agar plates could result in proliferation of bacterial cells in the liquid on the agar surface with a loss of distinct colonies.

After incubation at 24-30 C, usually from 6-14 days, the resulting bacterial colonies (Fig. 2-1) are counted. For counting purposes, dishes containing fungal or bacterial spreaders or large clear zones of antagonism should be discarded. The average number of colonies per dish is multiplied by the dilution factor to obtain the number per gram in the original soil sample. Single colonies

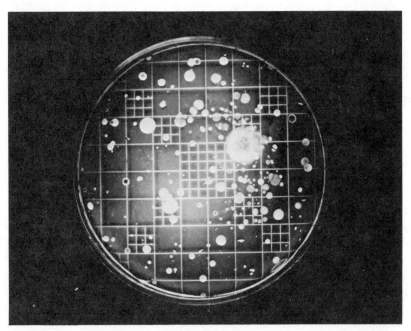

FIG. 2-1. Actinomycetes and bacteria (and one fungal colony) on a soil-dilution plate as viewed on a Quebec colony counter.

may be transferred to tubes of agar media for further study. Some difficulty may be encountered in distinguishing colonies of actinomycetes from bacterial colonies on dilution plates. In the following section on isolation of actinomycetes from soil, characteristic colony forms of each are listed which are helpful for identification. If difficulty is still encountered, pentachloronitrobenzene (PCNB) may be added to the isolation medium at a final concentration of 25-50 ppm. According to Farley (1967), this chemical retards the development of actinomycete colonies and has no effect on, or even favors development of bacterial colonies. It may be added to any of the common media listed at the end of this section.

a. **Use of dispersing agents and viscous materials.** Use of chemical dispersing agents has been recommended by several authors for the initial dilution. Higher counts of bacteria were obtained by Bhat and Shetty (1949) when 0.7% saline was used as the diluent instead of distilled or tap water. Damirgi et al. (1961) obtained highest numbers of bacteria from soils shaken for 25 min in a saline solution (0.2% NaCl) with 0.05% Na_2CO_3. Gamble et al. (1952) recommended use of a 0.5M solution of sodium hexametaphosphate for the initial dilution, while Hirte (1962) used a 0.1N solution. Jensen (1962) found that dilutions containing 0.5% sodium hexametaphosphate yielded higher numbers in some soils, but lower numbers in other soils. In repeated tests, we have found no distinct increases in bacterial counts in agricultural soils by using sodium hexametaphosphate (at levels of 0.2-3.0%) in the initial dilution with the dilution-plate procedure outlined above.

Hornby and Ullstrup (1965) recommended that soil should be suspended, for the primary dilution, in a viscous solution. Variation in the amount of soil samples pipetted from a solution containing 1% carboxymethycellulose or 0.2% agar was smaller than that from a water solution without these materials.

b. **High frequency vibration (Stevenson, 1958).** Treating the first dilution of a soil-plating series with high frequency vibrations resulted in initial increases in the numbers of bacteria, actinomycetes, and fungi appearing on the plates. A 4 min treatment of the first dilution on a 10-Kc Raytheon magnetostrictive oscillator resulted in a 350% increase in colonies of bacteria, a 200% increase in fungi, and a 20% increase in numbers of actinomycetes.

2. Membrane-filter technique (Davey and Wilde, 1955).

Molecular membrane filters are cellulose-plastic-porous membranes available in distinct pore-size grades ranging from 5 μ downward in pore size to 10 mμ. The 0.45-μ pore-size filter (available from the Millipore Filter Corp., Bedford, Mass.) has a variation in pore size of only ± 0.02 μ. Each square centimeter of the filter contains millions of capillary pores which occupy approximately 80% of the total filter area. The 0.45-μ pore-size filter prevents the passage of nearly all known bacteria. Entrained organisms are retained quantitatively from large volumes of fluid, and remain evenly distributed on the filter surface. Collected organisms may be stained *in situ* on the filter surface. They may also be grown into microcolonies or macrocolonies before staining and counting or before transferring to pure culture. For counting purposes the filter containing the colonies may be rendered completely transparent by filling its pores with immersion oil or mounting fluid having a refraction index of 1.5. The procedure is as follows:

Soil samples are passed through a sieve with openings of 841 μ. Water is added and the samples are incubated at field capacity for 48 hr at 23 C. One ml of the incubated material is taken with a calibrated scoop and dispersed in 199 ml of sterile distilled water in a Waring Blendor for 30 sec. Then the suspension is diluted in sterile water blanks until the desired dilution is reached. The final dilution will depend on the number of bacteria in the soil sample used and should be determined by trial.

A sterile membrane filter is placed in a previously sterilized special filter holder with attached flask (see Fig. 2-2), and a 100-ml aliquot of the suspension is drawn through with the help of an electric vacuum pump. The filter is transferred immediately to the surface of a nutrient agar medium in a petri dish or to an absorbent pad in a petri dish saturated with a nutrient solution. The solution will diffuse through the filter and permit development of colonies on its surface. After incubation at 24-30 C for 48 hr or longer, colonies on the filter can be counted, and transfers to pure culture can be made from the filter.

3. **Culture media for isolating soil bacteria.**

No single culture medium has been universally accepted for isolation of soil bacteria. The most widely accepted media are those containing soil extract. Below is a list of some of the media used (see Chapter 16):

a. **Media containing soil extract.** For most soils, higher numbers of species of bacteria are obtained on dilution plates with soil extract in the medium than on plates without soil extract. Generally, when isolating soil bacteria, the assumption is made that there is a direct positive relationship between numbers of bacterial colonies per plate and numbers of species obtained. James (1958) found that extract prepared by heating was better than cold water extract and that freshly prepared extract was better than extract prepared previously and dried. Egdell et al. (1960) found no evidence to support the suggestions that higher counts are obtained by the use of soil extract media containing extract prepared from the same soil as the sample tested. The source of soil appears to be unimportant so long as the soil for extract preparation is not of extreme type and has been well fertilized and cultivated.

Soil Extract Agar (Lochhead, 1940). This medium contains soil extract and K_2HPO_4. It has been used extensively as a nonselective basal medium for

FIG. 2-2. Membrane-filter apparatus with attached vacuum pump. (Courtesy Millipore Filter Corp.)

isolating a variety of soil bacteria used in studies on classification of these organisms based on nutritional requirements.

Soil Extract Agar (Allen, 1957). This is similar to that of Lochhead, but contains 1 g of glucose per liter.

Soil Extract Agar (Bunt and Rovira, 1955). The medium consists of soil extract, yeast extract, peptone, and mineral nutrients.

Soil Extract Agar (James, 1958). According to James, the method of preparation of soil extract is important. Lower counts of bacteria were obtained on dilution plates prepared with "diluted" soil extract than with "full-strength" soil extract.

    b. **Media without soil extract.** A number of media have been proposed consisting of standard pure materials found in most laboratories. Use of these media presumably could favor a better comparison of results of data obtained at different times or in different laboratories.

Sodium Albuminate Agar (Waksman and Fred, 1922).
Thornton's Standardized Agar (Thornton, 1922).
Modified Hutchinson's Agar (Bhat and Shetty, 1949).

B. ISOLATION OF ACTINOMYCETES FROM SOIL

Actinomycetes are usually isolated from soil by the dilution-plate method, but the filter-membrane technique has also been used with success. Procedures are similar to those used for isolating bacteria, except that in most cases lower final dilutions are plated.

It is easier to distinguish between bacterial and actinomycete colonies on dilution plates if the colonies of both develop on the agar surface. The procedure for obtaining surface colonies outlined in the dilution-plate method can be used. Porter et al. (1960) suggested the following modification which favors the development of microorganisms as surface colonies. A basal layer (15 ml) of medium is poured into each petri dish and left to harden. Soil samples are diluted, 1 ml of each desired dilution is pipetted into tubes with 14 ml of agar medium, and the tubes are maintained at 48 C in a water bath. Each tube is shaken and 5-ml quantities are pipetted onto the hardened surface of the agar in each petri dish. Composition of the agar of both basal and surface layers is identical.

Even when the above methods are used, some difficulty might be encountered in distinguishing between actinomycete and bacterial colonies. The following hints are useful for identifying colonies on dilution plates: (1) colonies that have a white, gray, or black powdery surface are mostly actinomycetes; (2) colonies, observed visually without a microscope, that have a distinct halo with a darker interior are mostly actinomycetes; (3) spreading colonies, or clear, glasslike colonies, are mostly bacteria; (4) colonies that have very shiny surfaces are mostly bacteria; (5) the lens- or lemon-shaped colonies are mostly bacteria; (6) the very thin surface colonies are mostly bacteria; and (7) actinomycete colonies usually are firmer than bacteria when the colonies are pricked with a sharp needle. Whenever there is doubt as to identity, the edge of the colony can be examined under the low power objective of a microscope. Actinomycete colonies (with very few exceptions) have radiating hyphae at the edge of the colony. Bacteria always have smooth colony edges, although some might be wavy or indistinct.

1. **Special soil-sample treatments for isolation of actinomycetes.**

    a. **Phenol treatment (Lawrence, 1956).** This procedure reduces bacterial and fungal contaminants. Two drops of a 1:20 soil dilution are added to 10 ml of a 1:140 dilution of phenol. After 10 min one drop of the phenol-soil dilution is placed in 12 ml of melted agar medium in a petri dish.

    b. **Centrifugation technique (Rehacek, 1959).** This technique was designed to be used where the numbers of actinomycetes (in relation to fungi and bacteria) are low, and where low dilutions are required. One-gram soil samples are ground with a mortar and pestle in a small amount of water prior to dilution.

Grinding is important, since much lower numbers of actinomycetes are obtained with samples not treated in this manner. Filaments of actinomycetes tend to stick to larger soil particles and settle out during centrifugation. The soil is then diluted with water to the desired final dilution. Samples of the final dilution may be centrifuged at 3000 rpm for 20 min, but speed and time of centrifugation should be determined by trial, since much variation exists in number of actinomycetes, number and kind of contaminating organisms present in different soils, and types of centrifuges available. Samples (1 ml each) of the supernatant are plated on agar media in petri dishes.

Most fungal spores and many large-spored or large-celled bacteria such as *Bacillus cereus, B. subtilis, B. mycoides,* and *B. mesentericus,* which commonly occur on dilution plates, settle to the bottom of the centrifuge tubes.

2. Culture media for isolating soil actinomycetes (see Chapter 16).

a. **General purpose media.** Any of the media listed under "Culture media used for isolating soil bacteria" may be used for isolating actinomycetes.

b. **Selective media.** None of the media are entirely selective for all actinomycetes to the exclusion of all bacteria and fungi.

Benedict Agar (Porter et al., 1960). Actinomycete colonies are favored over bacteria by selection of the nitrogen source, L-arginine. If trouble with fungi is encountered, pimaricin at 50 µg/ml may be added to the medium.

Egg Albumin Agar with 40 µg/ml actidione (Corke and Chase, 1956). This medium was used to study actinomycetes in forest soils, where populations of fungi were high.

Glucose-Asparagine Agar with 0.4% sodium propionate (Crook et al., 1950). Growth of fungi is inhibited.

Glucose-Asparaginate Agar (Conn, 1921). Actinomycetes are favored over bacteria by selective utilization of the nitrogen source, sodium asparaginate.

Chitin Agar (Lingappa and Lockwood, 1962). This medium is selective for growth of actinomycetes and suppresses most bacteria and fungi. It was found that fewer actinomycete colonies developed on conventional media, such as Czapek's agar, glucose-asparagine agar, and egg albumin agar than on chitin agar.

Arginine-Glycerol-Salt Agar (El-Nakeeb and Lechevalier, 1963). This medium was compared with chitin agar, Benedict agar, glycerol-asparaginate agar, and others. It was superior with respect to numbers of colonies of actinomycetes obtained and supported the isolation of many types.

Starch-Casein Agar plus antifungal and antibacterial antibiotics (Williams and Davies, 1965). This medium was compared with chitin agar, Benedict agar, arginine-glycerol-salt agar, and egg albumin agar media. Starch-casein agar plus the antifungal antibiotics and chitin agar were superior to the others with respect to numbers of actinomycete colonies per plate. Colonies on the starch-casein medium were easier to count and such plates contained fewer fungi. Slightly fewer actinomycete colonies per plate were obtained when antibacterial antibiotics were added, but much higher ratios of actinomycetes to bacteria were obtained.

C. ISOLATION OF FUNGI FROM SOIL

Hyphae of some fungi fragment more readily than others during processing of soil, and some fungi sporulate more profusely than others. Therefore, techniques for their isolation are unsatisfactory from the standpoint of obtaining quantitative data. Also, qualitative data are limited, in that only fungi that will grow on the medium or substrate used are isolated. It has been suggested that a more accurate ecological picture of fungi in soil might be obtained by using two or more methods simultaneously, along with direct microscopic examination of the soil being assayed.

1. Dilution-plate method.

With this commonly used method a variety of fungi are obtained, but species that sporulate profusely are most often selected. Information obtained can be misleading, in that many fungi that occur in soil primarily in the mycelial condition are rarely isolated, yet the volume of space occupied and their metabolic activity may be as great or greater than the often isolated sporing species. The dilution-plate method is useful, however, when a large number of different kinds of soil fungi are desired, when studies deal with the relative response of soil fungi to various soil treatments, or when comparisons of methods for isolating soil fungi are being made.

The procedure for isolating soil fungi by the dilution-plate method is outlined under "Isolation of bacteria from soil." Final dilutions should be determined by trial, but they are usually lower than those required for bacteria and actinomycetes. For many soils, 1 ml of a 1:10,000 dilution placed in each dish yields a satisfactory number of fungal colonies (Fig. 2-3). Paharia and Kommedahl (1954) recommended that 1 ml of the final dilution be distributed over the surface of a solidified agar medium which has been poured into dishes 2-3 days previously. More fungal colonies per dish were obtained in this way than when the soil suspension was incorporated into the medium at time of pouring.

A "dropper plate" modification was proposed by Schenck and Curl (1961) to improve the dilution-plate technique. Since the dilution method is lengthy and requires considerable glassware, this shorter procedure speeds up assaying large numbers of soil samples. It consists simply of agitating 50 mg of soil (dry weight basis) for 20 sec in 20 ml of sterile water in a 1-oz screw-cap bottle. Two drops of the mixture are placed with a medicine dropper in each petri dish and swirled in 15 ml of agar medium.

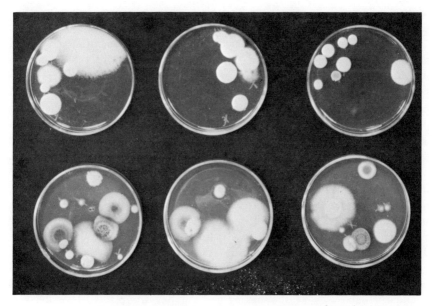

FIG. 2-3. Fungi on soil-dilution plates after 7 days of incubation at 24 C. Medium is Peptone-dextrose-rose bengal agar with 2 µg/ml of Aureomycin.

Another shorter procedure, the syringe technique, in which the actual weight of soil placed in each dish is calculated, was proposed by Rodriguez-Kabana (1967). A 25-g equivalent of oven-dry soil and a Teflon-coated magnetic bar are placed in a 500-ml aspirator bottle with 225 ml of sterile water. The bottle, with tubing outlet covered with a rubber cap, is placed on a magnetic stirrer for 3 min and the soil suspension is sampled while in motion with a 14-gauge cannula on a syringe. Approximately 2 ml of suspension are drawn into the syringe and individual drops are transferred into petri dishes. Agar medium is poured into the dishes and the procedure is followed in the usual manner. The amount of soil delivered in each drop of suspension is determined by drying and weighing 10 drops in each of 4 tared weighing cups. This method eliminates much of the error common in serial dilutions and pipetting, and allows a high degree of reproducibility and possible standardization.

2. The soil-plate method (Warcup, 1950).

This method, adapted from the direct-inoculation technique of Waksman (1916), was designed to study the ecological distribution of various species of fungi in soil. The soil-plate method permits isolation of fungi embedded in humus or attached to mineral particles that might be discarded in the residue when dilution plates are prepared. Spore masses tend to remain more intact in soil plates than in dilution plates, thereby favoring more valid data on the distribution of fungi in soil. Many limitations are common to both methods and they present essentially the same ecological picture of the fungal flora of the soil, but a greater number of species of fungi are usually isolated on soil plates than on dilution plates. Several Basidiomycetes occasionally appear on soil plates. Root pathogens and medium- to fast-growing fungi present in soil in relatively low numbers appear more often on soil plates than on dilution plates. The procedure is easier to follow, since it is less time-consuming and requires fewer laboratory materials than the dilution-plate method.

The amount of soil used to prepare a soil plate varies with the soil investigated, and is determined by trial. With many natural surface soils about 0.005-0.015 g results in a convenient number of colonies in each dish. A microspatula, made by flattening the end of a nichrome needle, is useful for transferring the microsample to petri dishes and for crushing soil aggregates. There is small variation in weight among transfers from the same soil sample.

Eight to 10 ml of melted and cooled agar medium is added, and the soil particles are dispersed throughout the agar. With sandy soils, adequate dispersal may be obtained by gently shaking and rotating the plate before the agar solidifies. If the soil is very dry, or contains a high proportion of clay, it is preferable to mix the soil particles with a drop of sterile water in the dish before the medium is added.

It is suggested that for purposes of studying the ecological distribution of fungi in soil, presence or absence of species in each soil plate be recorded, without regard to the number of colonies per plate.

a. **Plate-quadrat modification (Aberdeen, 1955).** This modification was designed to yield a more accurate estimation of the ecological distribution of fungi in soil, based on the presence or absence of a species in a quadrat of known size. It tends to eliminate the variability in soil-sample sizes on Warcup's microspatula. Holes of varying diameter are bored through brass plates of different thickness. The plate containing the hole of the desired volume (usually 0.5 mm^3 or more) is placed on a flat surface and a block of soil pressed firmly into the hole, the soil block being disturbed as little as possible. The soil is then smoothed off level with the surface of the plate. The soil sample in the hole (a quadrat of soil) is transferred to a petri dish by pushing the sample through the hole with a sterile needle. Warcup's procedure is then followed. For soils having large populations of fungi it may be necessary to suspend the quadrat in a weak agar solution and then transfer it to a number of dishes. Presence or absence of a particular species is recorded for each quadrat, and not for individual dishes.

Frequency percentages are used to estimate the average size and density of the colonies, or the volume of soil occupied, for a particular species of fungus. A species-volume curve is used to give an estimate of the extent of aggregation for the entire community of fungi present in a soil sample. For a discussion of the mathematical treatments involved, see Aberdeen (1955).

b. **Dilution with sand (Johnson and Manka, 1961).** This modification is useful when soils of high fungal populations are assayed. With the original Warcup procedure 5-15 mg of some soils, especially cultivated ones, may yield such large numbers of colonies per soil plate that they are difficult to count, identify, or transfer. This modification reduces the number of colonies per plate without losing many of the advantages of Warcup's soil-plate method. One-gram samples of soil are placed in wide-mouth 250-ml Erlenmeyer flasks containing weighed amounts of sand (24 g, 49 g, etc.) previously sterilized in an autoclave. The flasks containing the soil-sand mixtures are then rotated gently for 1 min to disperse the soil particles in the sand. Microspatula samples of this mixture are transferred to petri dishes, and soil plates are made in the usual manner. Data can be taken on the presence or absence of certain fungal species per plate.

If desired, the weight of soil on one microspatula can be estimated. Fifty or more spatula samples on the soil-sand dilution used are combined and weighed accurately. The average weight per spatula is calculated. The weight of soil on each spatula (W) can be calculated by the following formula : $W = \frac{XY}{Z}$, where X is the original weight of soil in each flask, Y is the average weight of soil and sand on each spatula, and Z is the original total weight of soil and sand in each flask. Thus, if dilution of soil to sand of 1:24 is made, and if the average weight of soil and sand on each spatula is 10 mg, then $W = \frac{1 \times 0.01}{25}$ or .0004 g. With this information the number of colonies or particular species of fungi per gram of soil can be estimated.

The method outlined above will not yield data on the spatial distribution of species of fungi in microquadrats of soil, but it has certain advantages over the standard water-soil dilution technique. It is easier and less time-consuming. In addition, heavier soil particles or aggregates containing fungal mycelia are plated; these tend to settle and be lost in water dilutions. Spore masses remain more intact; thus, species producing large numbers of spores form fewer colonies on modified soil plates than on water-dilution plates.

3. **"Immersion" techniques.**

Several procedures have been developed which consist essentially of placing agar media or other substances in soil and permitting growth onto the medium of fungi from specific loci in soil. The apparatus holding the medium is removed, and fungi colonizing the medium are isolated in the laboratory. The devices used are essentially modifications of the Immersion Tube (Chesters, 1948) or the Screened-Immersion plate (Thornton, 1952).

a. **Plastic disc (Wood and Wilcoxon, 1960).** Units are assembled with petri dishes and plastic discs (a fiberglass product of Filon Plastics Co., Hawthorne, California) 8.5 cm in diameter, and 1 mm thick, with 12 holes each 2 mm in diameter (Fig. 2-4). The discs are constructed with a band saw and drill press. About 25 ml of sterile water agar are poured into each petri dish, the discs (sterilized by autoclaving) are placed on the agar surfaces, and the dishes are covered with the petri dish covers until used.

To isolate fungi from soil, the covers are removed and the dishes pressed against a soil profile so that soil and disc are in contact. When the dishes are in place they are covered with a piece of aluminum foil and then with soil. The foil helps keep unwanted soil particles out of the dishes when they are removed from the soil. After being in contact with soil 2 or more days, the dishes are removed and the fungi are isolated. In the laboratory the positions of the holes in the discs are marked on the bottom of the dishes and then the discs are removed. Fungi in these areas are transferred.

b. **Profile plate (Anderson and Huber, 1965).** To study distribution and association of the actively growing fungi over a larger area, the profile-plate technique was developed. These plates are autoclavable polypropylene plastic

FIG. 2-4. Plastic disc containing holes, and petri dish with agar medium. (Wood and Wilcoxon, 1960. Plant Disease Reporter 44:594.)

plates (20 × 30 × 1.5 cm each) with 0.5-cm holes (1 cm deep) spaced at 2.5-cm intervals both vertically (7 holes) and horizontally (11 holes). They are cleaned, wrapped in aluminum foil, and autoclaved. Holes in each plate are filled aseptically with sterilized agar medium and each hole is covered with autoclaved plastic electricians' tape. The plates are then rewrapped in the aluminum foil and placed separately in large envelopes for transport to field sites.

A soil profile is prepared by driving a sharpened steel plate into the soil at right angles to the surface. Soil is removed from one side of the plate with a spade and then the plate is removed gently with minimum disturbance to the profile. One hole is punched in the center of each tape over each hole in the profile plate with a small sterile needle, and immediately the plate is placed firmly against the flat surface of the soil profile. Rows of holes are positioned vertically as desired on the profile. The soil removed to expose the profile is replaced and packed firmly against the back of the plate; the aluminum foil is placed over the plate to exclude soil and water.

After an appropriate period of exposure (usually 5 days), plates are removed, the tapes are peeled off in the laboratory, and fungi isolated by placing the agar plugs on agar medium in petri dishes.

    c.   **Immersion Tube (Mueller and Durrell, 1957).** Immersion tubes are prepared from plastic centrifuge tubes (Super-Dylon 1210-1HH made by the Dynalab Corporation of Rochester, N. Y.). Holes 3/16 inch in diameter are bored with a templet in a spiral arrangement through the walls of the tubes and then countersunk. Spacing of the holes may be varied as desired. The tubes are wrapped spirally with plastic electrical tape and filled to within 4 cm of the top with nutrient agar, plugged with cotton, and autoclaved.

In the field, a large needle is heated with an alcohol lamp or small blow torch and punched through the tape and the tube perforation into the agar. The needle is removed leaving an entrance for actively growing fungal hyphae.

After the tubes have been embedded in the soil for 4-6 days they are removed and taken to the laboratory, where the plastic tape is unwound, exposing one perforation at a time. A stiff transfer needle, the end flattened, can be used to transfer the agar and fungal invader to petri dishes containing rose bengal agar, or other suitable media.

    d.   **Capillary-immersion tube (MacWithey, 1957).** Capillary-soil-immersion tubes were developed for examination of soil microflora in small areas, such as pots and greenhouse bench plots. These tubes are made from pieces of 6-mm glass tubing about 8 cm long. The tubing is heated in the center and a capillary about 0.5-1.5 mm in diameter is drawn out. The capillary is broken off to a length of about 3-4 cm and the large open end plugged with cotton.

Forty to 50 capillary tubes are placed upright in a beaker and supported by a wire mesh. The beaker is then filled with melted agar to a height of about 4-5 cm and autoclaved. Immersion tubes are removed from the solidified agar and the plugged ends are capped with parafilm. The capillary is immediately inserted into a pilot hole made in the soil.

Tubes are removed from the soil periodically, cleaned, and examined under the microscope for the presence of fungal mycelium within the capillary. Colonized tubes are surface sterilized and the ends are broken off and transferred to agar plates. Two or three days later, mycelium growing from the capillary may be reisolated and identified.

 e. **Strip bait (Luttrell, 1967).** The strip bait consists of filter paper discs sealed between two strips of plastic electricians' tape. The discs are spaced at appropriate intervals along the strip beneath perforations in the tape. Preparation of the strips is as follows: Whatman No. 3 filter paper is soaked in the desired nutrient solution, then air dried on screens. Discs, 10 mm in diameter, are cut from the dried paper and placed over 5-mm holes cut in 20-mm-wide electricians' tape spaced 25 mm apart. Five discs are placed over each hole in the tape. An identical strip of tape with punched holes is placed over the stacked discs so that the holes are centered over the discs. The tapes are pressed firmly together and around the discs. One end of the tapes is left with the tips folded under so that they can be easily pulled apart. Code notches may be cut in the margins of the tapes to indicate composition of the nutrient solution used. Prepared tapes are then placed in paper packets, autoclaved, and stored until needed.

Sterile tapes, 23 cm long, are placed adjacent to a soil profile and positioned as desired. Soil removed during digging is replaced against one surface of the strip. After an exposure interval, the tapes are returned to the laboratory and the stacks of filter paper are exposed in sequence by pulling the tapes apart. From the middle disc in each stack, tiny pieces of filter paper are removed with sterilized forceps and placed on nutrient agar in petri dishes.

A number of genera of fungi can be isolated from strip baits. An advantage appears to be that strip baits soon assume the approximate moisture level of the surrounding soil. They can be left in soil for more than a month for studies of succession of fungal species.

4. Methods for isolating fungal hyphae from soil.

 a. **Physical dissection (Warcup, 1955).** This technique was developed as a result of the observation by Warcup that, when a soil suspension is prepared, many of the fungal hyphae remain with the heavier soil particles of the residue. Removal of the fine suspended material from the residue permits visual examination of the latter for the presence of individual hyphae or hyphal masses, which may then be removed and grown on agar media. This method permits the isolation of fungi that rarely appear on dilution plates.

A soil crumb of about 1.0-1.5 g is placed in water in a beaker and left to become saturated. After 4 or 5 min the crumb is broken apart by filling the beaker with a rapid jet of tap water. The heavier soil particles are allowed to settle for 1.0-1.5 min and then most of the suspension is poured off. More water is added, the heavier particles are allowed to settle, and the supernatant fluid again is removed. This procedure is continued until the liquid remains clear after standing for 1.0-1.5 min.

The soil particles of the residue are then distributed in a small quantity of water among three sterile petri dishes. The material is examined with a binocular dissecting microscope for the presence of fungal hyphae. Individual hyphae, or portions of hyphal masses, are removed with fine forceps from among the soil particles and are placed in a drop of sterile water in a clean, sterile petri dish. Hyphae attached to mineral grains or to humus particles must be separated carefully from the attached particle. Hyphal masses should be teased out, since they usually consist of more than one species of fungi.

When sufficient hyphae are in the dish, 10-15 ml of cooled agar medium are added. The hyphae are dispersed by shaking and rotating the dish before the agar solidifies. Then the hyphae are located, ringed, and numbered on the bottom of the dish. The dishes are examined daily under a microscope (120X) for fungal growth. Tips of growing hyphae that are not contaminated are cut out in agar blocks and transferred to fresh media. Careful examination is necessary to determine whether the colonies grow from hyphal strands or from humus particles, since spores are likely to be attached to humus particles. Colonies developing from spores or humus particles should be cut out to prevent them from overgrowing ungerminated spores or slowly growing hyphae. Dishes should be examined periodically up to 10 days after the agar medium is added. Czapek-Dox agar plus 0.5% yeast extract (pH 5.6-5.8) diluted to 1/6 normal strength is recommended as the isolation medium (Warcup, 1950).

b. **Nylon strip (Gams, 1959).** Strips of nylon gauze (about 2 cm wide) are cut so that they are slightly longer than microscope slides. Parallel cuts (1 mm apart) are made lengthwise in each piece of gauze but each end is left connected. Strips thus prepared are placed on slides, and the uncut ends are folded under the slides and fastened with cellophane tape. They are buried in soil for 1 week, then removed, rinsed with sterile water, and examined for hyphae caught in the meshes. The strips are separated with a sterile knife and placed on the surface of nutrient agar containing 3 µg/ml Aureomycin (chlortetracycline) in petri dishes. Developing colonies can be isolated and transferred to pure culture.

c. **Soil washing (Watson, 1960).** This technique is based on the finding by Warcup (1955) that hyphae tend to remain closely associated with soil particles in a soil-in-water suspension. A large percentage of isolates originating from such hyphae can be obtained by washing the soil to remove most of the fungal spores.

Hyphae that are attached to the residual soil particles are left in the residue from which dilution or soil plates are made. For details of this technique, see Chapter 5.

The following soil-washing apparatus can be used for large scale, repetitive experiments (Williams et al., 1965).

The apparatus (Fig. 2-5) consists of 1-6 washing boxes made from perspex (plexiglas). Each box (13 X 4 X 4 cm) is fitted with a stack of 3 stainless steel sieves (mesh size 1.0, 0.5, and 0.25 mm), the sieve of largest mesh size being uppermost. Each box has a removable frontpiece and is fitted with 4 perspex tubes. Sterile water is supplied to the boxes from a tank (A) by means of rubber tubes (B) which run into glass measuring bulbs (C). The bulbs are fitted with side arms that are plugged with sterile cotton to prevent contamination during the washing procedure. With the graduated bulbs, a measured volume of sterile water can be quickly introduced into each box. Water is removed from the boxes through rubber tubes (E) leading to a common outlet pipe (F). Compressed air, after passing through a cotton filter, enters the apparatus through a common glass tube (G) fitted with side tubes (H), supplying each box. Air passes out of the boxes through rubber tubes (D) which are fitted with terminal air filters. All rubber tubing is fitted with screw clamps to facilitate control of the flow of air and water.

Rubber tubing, measuring bulbs, and perspex boxes, sealed and fitted with their sieves, are sterilized with 95% alcohol. The other components can be autoclaved. A suitable volume of sterile water is introduced into the reservoir tank. A quantity of soil (3-4 g) is weighed and placed on the top sieve of each box to be used. The frontpieces are replaced, their edges sealed with plasticene and secured by two perspex bars bolted across the front of the boxes. A suitable volume of sterile water (40 ml) is measured into the glass bulbs and run into each box. The compressed air supply is then turned on and the clamps controlling the air inlets to each box opened until all boxes are receiving a supply of air, which causes vigorous, controlled agitation of the soil suspensions. This agitation is continued for 20 min and during this time the measuring bulbs can be refilled with water. The air supply is gradually reduced and the inlets to each box closed. Water is then quickly removed from the boxes by opening the clamps controlling the outlet tubes. This process can be repeated 30 times in 90-100 min.

After washing soil 25 or more times, soil particles from the different fraction-sizes are plated. For comparison, dilution plates and soil plates (Warcup's method) of unwashed soil may be made. If desired, individual particles can be plated by rinsing the sieves into dishes of sterile water. Discrete soil particles are picked out with forceps and placed individually on the surface of agar medium.

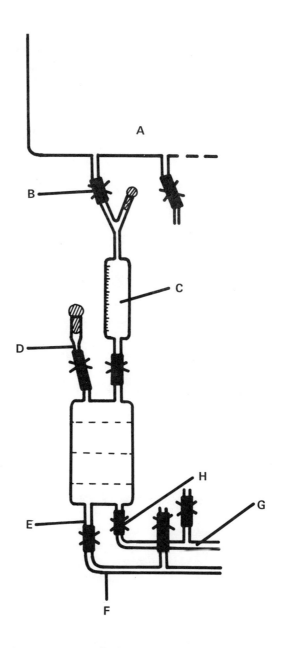

FIG. 2-5. Soil-washing apparatus. (Williams et al., 1965. Plant and Soil 22:167-186.)

Microorganisms from soil 23

5. Selection of an isolation method.

Direct evidence has been obtained that most colonies of fungi in soil-dilution plates develop from spores (Warcup, 1955b; Hack, 1957). Williams (1967), however, found that only 17% of the fungal colonies originated from spores, and 60% originated from humus particles. If the results reflect sporing activity, this would be a major disadvantage of the dilution-plate method. In defense of the method, Montegut (1960) proposed that (1) a large number of fungal species do not exhibit in the soil a frequency proportional to their sporulative capacity on artificial media, (2) sporulation reflects previous mycelial activity, and (3) the dilution-plate method can be precise when standardized, and data obtained can be analyzed easily by conventional statistical techniques. The method, therefore, would be useful especially for world-wide analysis of the microflora in different horizons of a given soil, for the determination of frequency of occurrence of species and its variations with ecological factors, and for other comparative studies.

When comparisons were made of the dilution plate and Warcup's soil-plate methods, it was found by Warcup (1950) and Agnihothrudu (1962) that the soil-plate method was superior with respect to isolation of greater numbers of species. Manka et al. (1961) found a greater number and variety of species on soil plates made from soil previously diluted with sand than on plates made from soil without dilution. Despite careful handling to ensure that the soil suspension is in motion while pipetting soil-water dilution series, larger soil particles tend to be left in the water blanks, and fungal hyphae normally attached to these particles are not isolated. The soil-plate method tends to give a more accurate picture of the ecological distribution of fungi in soil since heads of spores are not broken up as readily as they might be in water dilutions. Moreover, the soil-plate method is less time-consuming and does not require as much glassware as the dilution-plate method.

Due to the nature of the immersion-plate or tube methods, inherent difficulties exist when attempting to compare their effectiveness with those of the dilution- or soil-plate methods. The immersion plates and tubes tend to select fungi capable of growing out of soil (under conditions existing in the soil at the time of isolation) through an air space and onto an agar surface. The speed with which they grow under these specific conditions and the competition among the members present in the microhabitat appear to influence the types obtained on immersion plates or tubes. Despite these inherent difficulties of comparison, Chesters and Thornton (1956) made parallel studies of unlike techniques for isolating soil fungi. They concluded that isolations with screened-immersion plates exhibited a wider range and variety of species than those obtained by other methods (immersion tube, direct inoculation, modified Rossi-Cholodny buried slide, dilution plate, and soil plate), particularly for species of *Mortierella* and dematiaceous fungi, and also provided a quantitative measure of the relative distribution of species.

An accurate, complete, and unbiased picture of the distribution of fungi in soil cannot be obtained by using any one of the methods previously described. For critical ecological studies it is suggested that two or more unlike methods be employed simultaneously, along with direct-observation techniques (Chapter 3). For example, the immersion tube, the soil-plate, and the direct-microscopic-examination method of Jones and Mollison could be used advantageously together. Fast-growing fungi such as *Mortierella* and *Mucor* are most often isolated with immersion tubes. Also, *Rhizoctonia*, *Fusarium*, and *Pythium* occur frequently in immersion-tube isolations. Medium- to fast-growing species such as *Penicillium*, *Trichoderma*, *Aspergillus*, *Fusarium*, and *Mortierella* are most frequently isolated on soil plates. If soil is diluted with sand prior to plating so that fewer colonies develop per plate, many slow-growing fungi such as *Trichocladium*, *Talaromyces*, *Torula*, *Sporotrichum*, and *Periconia* may also be obtained (Johnson, 1964). As a supplement to these two techniques, an estimate of the number of spores and the total length of mycelium occurring in the soil being assayed can be obtained with the direct-microscopic-examination technique of Jones and Mollison. When comparing the effects of various soil treatments on fungi, one method of isolation alone might be sufficient, but the user must remember the limitations of the method selected and take them into account when discussing his results.

The number and kinds of fungi isolated will depend not only on the soil and isolation technique selected, but also on the agar medium used, length of the incubation period, and temperature and pH of the medium during incubation.

6. Agar media for isolating soil fungi.

Selection of an agar medium depends partly on the isolation method and partly on individual preference. Some studies have been made to compare the efficiency of different kinds of media and several conclusions have been reached:
a. Only fungi that will grow on the medium selected are isolated.
b. Antimicrobial agents (antibiotics, rose bengal, sodium propionate, etc.) should be incorporated into the medium, especially when the dilution-plate or soil-plate methods are employed. Although equally effective for inhibiting bacteria, some agents are more efficient than others for restricting the size of spreading colonies of fungi and result in larger numbers of fungal colonies isolated.
c. Certain environmental conditions during the incubation period might affect the efficiency of bacterial inhibitors, e.g., growth of fungi is inhibited on media containing rose bengal previously exposed to sunlight (Pady et al., 1960).

d. With few exceptions (James, 1959), there is no distinct advantage in using soil extract in the medium. Soil extract incorporated in the isolation medium tends to overcome the beneficial colony size-restricting effect of rose bengal and does not increase the total numbers isolated (Sewell, 1959; Johnson and Manka, 1961).
e. One medium, effective for isolation of fungi from one soil, might not be as effective with another soil.
f. Fungi are easier to identify on some isolation media than on others.
g. Certain genera, existing in soil in low numbers, may be isolated by using baits of various kinds in the medium.

Below is a list of some of the most commonly used agar media (ingredients in Chapter 16). Although some of the media may be somewhat better than others with certain soils, all (except water agar) will yield a maximum of fungal species. Not included are media containing acid, since it has been shown that acid media are generally inferior with respect to numbers and kinds of fungi isolated from soil.

a. Dextrose-peptone-yeast extract agar (DPYA) with sodium propionate, oxgall, and antibiotics (Papavizas and Davey, 1959a).
b. Ohio agar (OAES agar) with sodium propionate, oxgall, and antibiotics (Schmitthenner and Williams, 1958).
c. Peptone-dextrose-rose bengal agar with 30 µg/ml of streptomycin (Martin, 1950) or with 2 µg/ml of aureomycin (Johnson, 1957).
d. Peptone-dextrose agar with 500 ppm of Phosfon and 30 µg/ml of streptomycin (Curl, 1968).
e. Potato-dextrose agar with 100 µg/ml of novobiocin (Butler and Hine, 1958).
f. Potato-dextrose agar with 100 ppm of streptomycin and 1000 ppm of a nonionic surfactant, such as phenyl polyethylene glycol ether with concentration of 7 (NP-27) or 10.5 (NPX) moles of ethylene oxide (Steiner and Watson, 1965).
g. Water agar (2% agar). This is used extensively in immersion plates and immersion tubes. Antibiotics, such as streptomycin, are sometimes added to reduce bacterial contamination.

D. TECHNIQUES FOR ISOLATING SPECIFIC GROUPS OF FUNGI FROM SOIL

Some methods are designed to select out specific groups of the soil microflora not ordinarily isolated with the usual techniques. Use of some of these methods along with dilution plates or soil plates, that yield a large variety of soil fungi, might produce a more complete ecological picture of the soil microflora. Plant pathologists may find some of these methods useful, especially those concerned with isolating chytrids, mycorrhizal fungi, and nematode-trapping fungi. Some of the other groups are of interest to plant pathologists since they are involved in decomposition of organic matter and may affect the activities of root-disease fungi.

1. Acrasiales (Kitzke, 1952).

Escherichia coli is utilized in this method as the food supply for these organisms. Petri dishes containing nutrient agar are smear-inoculated with the bacterium to produce a dense growth. After 24 hr of incubation the cells are skimmed off the agar surface and suspended in sterile water. Soil samples to be tested are placed in 25-ml specimen bottles so that each bottle is about one-half full. The soil in each bottle is sprinkled liberally with the bacterial suspension, enough to moisten the soil but not inundating it. The bottles are covered loosely with their caps and incubated at 20-23 C. If present in the soil samples, pseudoplasmodia will be visible in about 2 days, and after 3 days numerous fruiting bodies will appear on the soil crumbs. If desired, subculturing from the isolation vials may be accomplished by transferring a sorus to agar containing a 24-hr-old streak of *E. coli*.

2. Chytrids (Willoughby, 1956).

Chytrids can be isolated from soil by use of baits such as cellophane, grass leaves, or onion scale epidermis. Grass leaves (such as *Poa annua*) are bleached in absolute alcohol, which is then removed by transferring the leaves to boiling water. Strips of the epidermis from succulent bulb scales of onion are removed and boiled in water to leach out most of the soluble substances. All of the baits are autoclaved in distilled water before use.

Three g of soil are placed in petri dishes and barely covered with charcoal water or soil extract solution. The baits are introduced and the dishes are left at room temperature. Charcoal water (Couch, 1939) is prepared by boiling 2 g of powdered animal charcoal with 500 ml of distilled water for 1 hr. The solution is filtered and autoclaved. Soil extract solution (Hanson, 1945) is prepared by placing 280 g of soil in 1 liter of distilled water for 2 days, followed by filtration and autoclaving. If chytrids are present in the soil samples they will appear within 2 days on the baits, which may then be removed for microscopic examination. By mounting the baits in fresh charcoal water, sporangial dehiscence often can be induced.

3. Saprolegniales and Pythiales (Harvey, 1925; Dick and Newby, 1961).

Soil samples (about 10 ml each) are placed in petri dishes and sterile, distilled water added so that the bait used can rest on the soil and at the same time be partially exposed to the atmosphere. After the water above the soil is left to clear for a short time, 3 autoclaved hemp seeds are placed in each dish. The dishes are incubated at room temperature near a north window. After 5-8 days a faint halo of fungal threads may be observed on the hemp seeds. The seeds are rinsed in sterile, distilled water to remove soil particles and placed into a fresh sterile petri dish half-filled with sterile, distilled water. After 2 days of incubation, identification of genera present can be begun. When species determination is desired, a unihyphal isolate may be grown on corn meal agar for further study.

4. Ascomycetes (Warcup, 1951; Warcup and Baker, 1963).

The proportion of ascosporic to nonascosporic species of fungi developing on soil plates can be increased if the soil is first treated with steam. Heat treatment stimulates dormant spores in soil to germinate. Many Ascomycete spores in soil are dormant and will not germinate and produce colonies on dilution or soil plates unless the soil sample is previously heated.

Samples of soil (125 g each) are placed in glass tumblers. It is assumed that soil type, structure, and moisture content might have an effect on the selective action of steam. Therefore, in order to obtain the maximum number of Ascomycetes, samples should be treated in a steamer at different time intervals, e.g., 0, 2, 4, 6, and 8 min. After treatment, the surface soil is removed to a depth of 1 cm, and a number of soil plates (Section C2) are prepared with the soil in each tumbler.

For a more critical evaluation of heat treatment, soil samples can be diluted with water to 1:100 and heated in a water bath. Treatment of the suspension at 60 C for 30 min has been successful. After treatment, dilution plates (Section A) are prepared and colonies of fungi obtained are examined.

It has been found that many ascosporic fungi survive treatment with ethyl alcohol. To reduce crowding of colonies on dilution plates, 2.5 g of soil are immersed in 60% alcohol for 6-8 min, then soil and alcohol are added to a water blank so that a dilution of 1:100 (and an alcohol concentration of less than 1%) is obtained. The soil dilution then may be heat-treated and dilution plates made as described previously.

5. Yeasts.

a. **Dilution-plate technique.** The dilution-plate method, as outlined in Section A, can be used to isolate yeasts from soil. Yeast colonies can be identified as such under a dissecting microscope and transferred to pure culture. An agar medium containing 4% (w/v) glucose and 1% (w/v) peptone and acidified to pH 4.0 has been used with success by di Menna (1957). Miller and Webb (1954) recommend use of Potato-Dextrose Agar containing 0.003% rose bengal or 1% oxgall.

b. Enrichment technique (Hesseltine et al., 1952). One-liter Erlenmeyer flasks containing 250 ml of tap water and 2.5 ml of corn steep liquor are stoppered and autoclaved. After cooling, each flask is inoculated with 1 g of soil. Aureomycin is added to the mixture so that a final concentration of 100 μg/ml is obtained. The flasks are shaken to disperse soil particles and incubated at room temperature near a window. Daily examinations are made to observe growth as evidenced by the formation of surface films on the liquid, or by the production of gas and turbidity. Mold colonies that appear on the surface of the liquid are shaken so that they settle to the bottom of the flask. Samples are taken aseptically from the liquid and observed under the microscope. When the yeast population becomes sufficiently great to see 2 or 3 cells per each high power field, the enrichment flasks have reached a stage where dilution plates may be made. One ml of the final dilution is added to each petri dish. Nutrient agar to which Aureomycin has been added at the rate of 50 μg/ml is then poured into each dish. After incubation yeast colonies can be transferred to pure culture.

6. Basidiomycetes.

Basidiomycetes are occasionally isolated in conventional dilution plates and soil plates. It is probable that in many cases sterile mycelia isolated by these methods are basidiomycetes, since they rarely fruit in agar culture. Warcup's hyphal-isolation technique (see Section C4) is a method that has been proposed to selectively isolate these fungi. Warcup found that many of the fungi growing from hyphae picked from soil were basidiomycetes. For cultural techniques used to obtain fructifications, see Warcup (1959), Goos (1960), and Warcup and Talbot (1962).

A medium developed by Russell (1956) has been used with success for isolating basidiomycetes from wood pulp. Many wood-rotting basidiomycetes will grow on this medium, but other fungi produce only very restricted colonies. It is proposed for use in isolating basidiomycetes from soil. The medium contains o-phenylphenol which inhibits common soil fungi. See Chapter 16 for ingredients. Growth of bacteria can be inhibited by adding an appropriate antibiotic such as Aureomycin at a concentration of 10-50 mg/liter.

7. Mycorrhizae.

Good selective methods for isolating mycorrhizal fungi directly from soil are not available. Many of these fungi are normally associated with roots and are not often found free in soil. Levisohn (1955), however, has been able to isolate directly from soil certain fungi known to form ectotrophic mycorrhizal associations; this was done with Warcup's hyphal-isolation method (see Section C4) in which rhizomorphs are picked from the soil suspension with fine needles. The rhizomorphs are washed for approximately 10 sec in 0.1% HCl, then rinsed in sterile water and plated on nutrient media commonly used for culturing mycorrhizal fungi. Ohms (1957) used a flotation method for collecting spores of a phycomycetous mycorrhizal parasite from soil.

8. Nematode-trapping fungi.

Conventional methods for isolating soil fungi fail to yield many of the predacious ones, although they are usually abundant in soil. No doubt this is partly due to failure of many of the predacious species to grow in artificial culture unless their animal hosts are present. A few of the common species do grow and sporulate in pure culture, but they are difficult to identify as being predacious because trap mechanisms are not often produced in the absence of the animal host. The methods described below are based on the presence of both the predacious fungus and the host nematode in the soil.

a. **Soil-plate method (Duddington, 1955).** This adaptation of Warcup's soil-plate method has been used with success for isolating predacious fungi. About 1-2 g of soil are scattered on the bottom of a sterile petri dish and a weak maize meal agar medium (see Chapter 16 for ingredients), cooled near to solidification, is poured in. The soil is slightly mixed with the agar by gently tilting the plate. Mixing should not be thorough, because complete distribution of soil over the plate might make subsequent observation difficult. Plates thus prepared are incubated at 15-20 C for up to 2 months, with periodic observation under a dissecting microscope during this time. Subcultures can be made by transferring small bits of agar containing the fungus and nematodes to a dish of sterile maize meal agar.

Species of nematode-trapping fungi were isolated by Tolmsoff (1959) from soil on milk agar. The medium consists of 2 g of powdered skim milk and 20 g of agar per liter. One gram or less of soil is scattered over the surface of the medium, and plates thus prepared are incubated at 25 C.

b. **Agar-strip method (Klemmer and Nakano, 1964).** Basically, this method is similar to the agar-plate method of Duddington, but modified so that semiquantitative data can be obtained on numbers of nematode-trapping fungi in soil. The soil sample is mixed with an equal volume of water and this slurry is mixed in a Waring Blendor. A 0.5 ml sample is pipetted into a petri dish and 30 ml of cooled, but still liquid 3% water agar is added. The soil is dispersed in the agar by gentle rotation and shaking. Long narrow strips (3 mm X 60 mm) of the hardened soil-agar mixture are cut from the dishes by means of two scalpels tied together with blades parallel. Each strip is transferred to a sterile petri dish and is flooded with 30 ml of weak corn meal agar. Finally, 1-ml volumes of a water suspension of saprophytic nematodes are pipetted over the soil-agar strips.

After several days of incubation, trapping structures can be seen on hyphae growing from the soil-agar strips. Examinations should be made in areas close by and parallel to the soil-agar strips and before extensive branching of hyphae has occurred. It is generally possible to distinguish between hyphae originating from separate loci within the strips. Such hyphae bearing one or more trapping structures are each given a count of one. Counts per dish can be converted to counts per gram of soil.

   c.   **Agar-disc method** (Cooke, 1961). Discs of a weak corn meal agar medium (see Chapter 16 for ingredients), 1 cm in diameter and 3 mm thick, are cut from the medium in petri dishes and placed on microscope slides. Four such discs are arranged in a row on each slide. The slides are buried singly in 15-cm petri dishes containing soil so that the discs lie about 1 cm below the soil surface. At weekly intervals some of the slides are removed from the dishes and the disc surfaces are washed with a strong, fine jet of water from a wash bottle. The discs adhere to the slide, and can be examined immediately under the microscope for both trap-forming predators and endozoic parasites.

   As an adjunct to the agar-disc method, Faust and Pramer (1964) suggested the use of Janus green dye to stain nematodes and predators on the discs. The dye is used at a concentration of 0.01% in 0.2M sodium acetate-acetic acid buffer at pH 4.6, and added drop by drop to the test material. Fungal walls stain dark green to blue; trapping structures stain most intensively. Nematodes, captured and killed as a result of fungal activity, are bright yellow. This is of interest since heat-killed nematodes stain dark green and living nematodes do not stain. Contrast between the dark green fungal structures and bright yellow nematode carcasses is augmented by lactophenol, which reveals hyphae located within dead nematodes.

   d.   **The most-probable-number method (Eren and Pramer, 1965).** An estimate of the actual number of nematode-trapping fungi in a given volume of soil can be obtained by utilizing a dilution series of soil. A known quantity of soil is suspended and diluted serially in sterile water blanks. One-tenth-ml portions from 6 or more dilutions are added to the surface of 1.5% agar in petri dishes. Three or more dishes should be inoculated from the soil suspension of each dilution. The exact dilutions necessary to give positive results should be determined by preliminary trials. The dishes are incubated at 28 C for 3 days, after which nematodes, such as pure cultures of *Panagrellus redivivus*, are added to the surface of the agar in each dish. The dishes are reincubated and examined periodically at a magnification of 100X for organelles of capture, such as sticky knobs, sticky spores, adhesive networks, rings, etc. Supplementary additions of nematodes are made when necessary and final observations of presence or absence of organelles of capture in each dish are recorded after 3 weeks of incubation. The most probable number of nematode-trapping fungi in soil is determined by reference to probability tables such as those found in Halvorson and Ziegler (1933) or Buchanan and Fulmer (1928).

9. Cellulose-decomposing fungi.

a. **Use of filter paper (Charpentier, 1960).** A nutrient silica-gel medium in petri dishes is treated with a small volume of soil suspension (1:10-1:100). The hardened gel is then tightly covered with a sheet of sterile filter paper in contact with the gel and incubated at 28 C in a humid atmosphere for 2-3 weeks. The filter paper can be stripped from the medium and cellulose-decomposing bacteria and fungi isolated from the paper.

b. **Cellophane method (Tribe, 1957).** Sheets of cellophane (transparent cellulose film) are cut into pieces (approximately 1.0 × 0.5 cm) and boiled in a large volume of distilled water to dissolve out plasticizers. After washing in another volume of distilled water, they are placed singly on 2-cm square cover slips and the excess water is drained off. The cover slips are then buried vertically in pots of soil or in soil in the field. They are recovered and examined periodically, and permanent microscope slides are made by placing the cover slip with adherent cellophane into picronigrosin in lactophenol for several hours, followed by washing in lactophenol. The back of the cover slip is washed with water, and the cellophane side is mounted on a slide in lactophenol and sealed with colorless fingernail varnish. Before preservation the slips can be examined wet under a microscope, and fungal spores transferred with fine needles to nutrient agar. Also, cellophane pieces may be cut into small fragments and plated for observation of the growing mycelia; or they can be brushed free of surface mycelium, then fragmented and plated to obtain cultures of fungi which produce "rooting branches" in the cellophane.

c. **Media containing "filter paper" cellulose.** A cellulose medium can be prepared easily in the laboratory (Bose, 1963). Small strips of Whatman No. 1 filter paper (or powdered filter paper) are immersed in concentrated HCl in the ratio of 3 g per 100 ml for 3 hr at 25-27 C with occasional shaking. The mixture then is poured with stirring into an excess of distilled water. After standing for 2-3 hr, the supernatant liquor is poured off and the residue washed in a Buchner funnel until free from acid. Finally, it is washed with 95% ethanol and air-dried. The cellulose thus prepared is used at 1% concentration and a weighed quantity is placed in a mortar, soaked with a few ml of Czapek salts solution, and made into a fine paste with a pestle. The paste is then made up to the required volume with Czapek salts solution. Agar is added (final concentration, 1.5%) and the medium is dispensed in flasks and autoclaved.

Immediately before pouring soil plates (Warcup's soil-plate method, see Section C2) each flask is shaken to obtain a uniform suspension. Bacterial growth may be suppressed by adding rose bengal (1:30,000) before autoclaving and Aureomycin (2 µg/ml) to the medium before pouring into petri dishes. The medium sets to an opaque mass with cellulose uniformly distributed throughout. The dishes are incubated at 30 C and in 4-6 days cellulolytic fungi are readily recognized by the formation of clear transparent zones around the growing colonies. Another useful cellulose medium (ingredients in Chapter 16) has been described by Eggins and Pugh (1962).

10. Lignin-decomposing fungi.

An enrichment technique was developed by Henderson (1961) for isolating fungi that attack lignin-related compounds (vanillin or p-hydroxybenzaldehyde). Briefly, it consists of adding quantities of soil to a nutrient medium containing these compounds as the sole carbon source. After 2 weeks of incubation, dilutions are made and fungi isolated on agar plates.

E. ISOLATION OF ALGAE FROM SOIL

Little is known about the importance of soil algae and their effect on plant growth. Due to their photosynthetic activity, algae that live in the surface layer of the soil add to the supply of organic substances. Certain blue-green algae may contribute to the fertility of the soil by nitrogen fixation. Also, certain algae produce antibiotics or toxins that inhibit the growth of other algae and retard root development of higher plants (Flint, 1947). It is possible that algae affect the development and severity of plant diseases by producing substances inhibitory or stimulatory to root parasites, or substances that predispose plants to attack by root parasites. For purposes of counting and identification, it is useful to isolate and maintain algae in semipure (or uni-algal) cultures. Sometimes an accurate determination of their numbers in soil is difficult because the cells of many species are enclosed in a mucilaginous matrix. The procedures of two methods for isolating soil algae in uni-algal cultures are given below. Also, a method for obtaining pure cultures is included.

1. Dilution-frequency method (Allen, 1949).

The moisture content of soil samples for study is determined, and 10 g of soil (on dry-soil basis) are diluted with water through a series of 1:100, 1:1000, 1:10,000, and 1:100,000. One ml from each dilution is transferred to each of several bottles containing 50 g of white quartz sand and 20 ml of Bristol's sodium nitrate solution (see Chapter 16 for ingredients). The sand medium should be sterilized previously. The bottles are incubated for several weeks in a window area or other source of light, and the number of bottles in which growth occurs is recorded. The most probable number of algae per ml and the number of algae present in the soil are calculated with the dilution-frequency method (see Halvorson and Ziegler, 1933).

2. **Moist-soil-plate method (John, 1942; Willson and Forest, 1957).**
Like Warcup's soil-plate method for isolating fungi, this method facilitates an analysis of kinds of viable algae under conditions where the field community is but little disturbed. It is based on the premise that the composition and frequency of occurrence of different kinds of algae in microplots yield information relative to the distribution and characteristics of the algal flora in a given area. No attempt is made to determine the numbers of algal cells per gram since, as with the fungi, this information can be misleading.

Ten- to 20-g samples of soil obtained from single or composite field samples are placed in sterile petri dishes. If desired, undisturbed soil samples of petri dish size, from ½ to 1 cm thick, are taken from the field with a sterile spatula and placed immediately in the dishes. The soil is saturated with sterile, distilled water or with a balanced mineral solution such as Bristol's (Chapter 16). Liquid is added periodically to keep the soil moist, but not continually saturated. The cultures are incubated at 20-24 C under continual or intermittent light. Most algae grow satisfactorily when exposed to 100-500 ft-c of incandescent or fluorescent light.

Periodically, small quantities of soil obtained with forceps are placed in drops of water on glass slides and examined microscopically for the presence of algae. At least a month of incubation should elapse before final examinations are made, since many algae require long periods of development before they can be identified with certainty. If desired, examination for individual types or species can be made in ten 5-mm square subsamples in each petri dish. Frequency of presence of types is computed as a percentage of the total subsamples or dishes examined.

If uni-algal cultures are desired they can be obtained from the petri dish cultures. Small portions of soil containing algae are placed in sterile Bristol's solution in Erlenmeyer flasks with cotton plugs. These liquid cultures are subjected to the same laboratory conditions as the original petri dish cultures for 2-3 weeks, or until a green ring of algae appears around the meniscus of the solution. Small quantities of this material are placed in drops of water in sterile petri dishes. Sterile Bristol's agar (Bristol's solution plus 2% agar), melted and cooled to 45 C, is added to the dishes. The dishes are shaken gently and incubated under light. Isolated colonies that develop are transferred to agar slants in culture tubes.

3. **Use of phenol to obtain pure cultures (McDaniel et al., 1962).**

To 50 ml of a culture containing 0.3-0.8 ml of centrifuge-packed algal cells, 2 drops of ARKO detergent (Deko Chemical Co., Culver City, Calif.) are added. After shaking vigorously, 2 ml of the mixture are added to 8 ml of a 1% phenol solution and then centrifuged for 5 min. The supernatant is discarded and 10 ml of distilled water are added to the cells; the tube is shaken vigorously to disperse the cells. After centrifuging for another 5 min, the supernatant is discarded and small amounts of the residue are transferred to petri dishes containing nutrient solution plus 2% agar and 0.5% dextrose. To ensure relative dryness of the agar surface, which is important for distinct colony growth, dishes can be prepared 2-3 days previously. The dishes are sealed with tape and incubated on a table under light. If viable bacteria are present, they usually appear within 24 hr.

CHAPTER 3
ENUMERATION AND OBSERVATION OF MICROORGANISMS IN SITU

Population estimates with more direct techniques are usually larger than those obtained by counting colonies on dilution plates. About 100 times more bacteria and 6 times more fungi were estimated by direct microscopic examination of soil (Skinner et al., 1952). The two methods apparently were not complementary since different information about microbial fluctuations was obtained. One should be aware of the limitations of both techniques. Results with the dilution technique could be influenced by (1) aggregation of bacteria in clumps despite vigorous shaking of the initial suspension, (2) adsorption of bacteria on soil colloids, (3) competition of colonies on dilution plates, (4) selectivity of the agar medium, and (5) exclusion of obligate anaerobes. Limitations of the direct microscopic examination technique that might influence results include (1) affinity of stains for nonviable particles, (2) frequent failure to stain viable cells, and (3) difficulty in distinguishing between bacterial cells and actinomycete spores, and between fungal and actinomycete filaments. Recent studies with fluorescent vital dyes for staining living bacterial cells have increased the validity of direct microscopic examination. Newer methods, designed to determine total hyphal length per unit volume of soil, have added to our knowledge of the complex nature of the activity of soil fungi.

A. DIRECT MICROSCOPIC EXAMINATION OF SOIL (CONN, 1918)

The types and relative abundance of microorganisms in soil can be determined by direct microscopic examination. This method has yielded considerable quantitative information about soil microorganisms, but their spatial relations are disrupted with the method of preparation.

One g of soil (on an oven-dry soil basis) is transferred to a test tube containing 9 ml of a sterile 0.015% agar suspension. The mixture is shaken gently to obtain a uniform 1:10 suspension. One-tenth ml of the suspension is transferred with a sterile pipette to the center of a glass slide. A flamed and cooled wire loop is used to spread the drop uniformly over 4 cm^2 of the slide. A ruled piece of cardboard under the slide facilitates this operation.

After the suspension has dried in air, it is fixed by immersing in 0.1 N HCl for 1 min. The excess acid is washed off the slide immediately by brief immersion in water, and the slide is dried thoroughly on a flat surface over a boiling water bath. The material on the slide is stained over the water bath with a 1% aqueous solution of rose bengal or carbol erythrosin. The stain should remain on the material for 1 min, care being taken to prevent drying. After washing, the slide is air-dried.

The microorganisms in 25 microscopic fields are counted under an oil immersion objective and the average number per field is calculated. The diameter of the oil immersion field is determined with a stage micrometer, and the area is calculated in square centimeters. From this information the number of microscopic fields in 4 cm^2 can be determined, and the total number of microorganisms on the slide (i.e., the number in 0.1 ml of a 1:10 soil dilution) can be estimated. Estimates of total microbial populations in the original soil sample can be made. Bacterial cells are normally present in varying concentrations in commercial powdered agar (Harris, 1969). Cells on control slides containing an agar suspension without soil should also be counted.

1. Soil-smear modification (Nash et al., 1961).

This technique was designed to observe and study chlamydospores of *Fusarium* in soil. A small amount of the infested soil is mixed with water to make a slurry, which is poured on a glass microscope slide (1 or 2 drops at a time) in an amount sufficient to be covered by a large cover slip. Acid fuchsin stain is added to the slurry in amounts required to give a deep red stain to the spores. The smears are examined under the high power of a compound microscope.

2. Agar-plate modification (Papavizas, 1967b).

This technique was developed for observation of large-spored fungi such as *Fusarium solani, Helminthosporium sativum,* and *Thielaviopsis basicola* in infested soil. Suspensions of the test soil (10 g on oven-dry basis in 190 ml of 0.5% carboxymethylcellulose) are comminuted for 1 min at low blender speed (8,000 rpm). The carboxymethylcellulose (CMC-70L type, low viscosity) is desirable to increase viscosity of the diluent and reduce rate of settling. Thirty sec after blending, 1-ml samples are withdrawn from 2 cm below the liquid surface and pipetted onto the surface of solidified agar in petri dishes. The dishes are swirled to distribute the suspension, which is then held without dish covers for 15-20 min to permit absorption into the agar. The dilution will depend on the kind of soil used, usually 1:10 for sandy soils and loamy sands or 1:20 for loamy and clay soils.

Immediately after suspensions are absorbed by the agar, three drops of lacto-fuchsin (500 mg of acid fuchsin in 300 ml of 50% lactic acid) are added to the surface. A large No. 1 cover glass is applied and pressed gently onto the agar surface. Microscopic examination is made immediately with a X25 or X40 objective, and numbers of ungerminated propagules are recorded per microscopic field. For quantitative estimates, the total area examined per dish is calculated with an ocular micrometer and, the dilution factor being known, it is possible to estimate propagule numbers per gram of soil by multiplying numbers estimated per dish by the dilution factor.

3. Agar-film modification (Jones and Mollison, 1948; Thomas et al., 1965).

Variable amounts of soil (between 0.5 g and 4.0 g) are placed in 5 ml of water and ground in a mortar for 5 min. The supernatant is decanted into a collection container, 5 ml of water are added again to the residue, and the material is ground for 2 min. This process is repeated 4 or 5 times. Finally, all the soil particles are collected and 20-25 ml of suspension are obtained. Molten agar is added to give a final dilution of the original soil sample of approximately 1 g soil in 50 ml of 1.5% agar. Preliminary experiments may be required to determine the exact desired concentration of soil:liquid.

Immediately before preparation of the film, the suspension is thoroughly agitated and left to settle for 5-10 sec. Then a drop of the material is pipetted from approximately 1 cm below the surface and placed on the platform of a hemacytometer slide (depth 0.1 mm). The well of the slide is immediately covered with a coverslip and the enclosed suspension is left to solidify. Because of the viscosity of the agar suspension, agar films are often thicker than 0.1 mm. To minimize this error, a small square weight (approximately 5 g) cemented to the coverslip can be placed over the suspension held in the well of the slide. See Tuite (1969) for instructions on use of hemacytometer slides.

After the suspension has hardened on the slide, it is immersed in distilled water and the coverslip is removed. The film is floated off onto an ordinary microscope slide, where it is air-dried at room temperature. Dried films are stained for 1 hr in the following stain: aqueous phenol (5%), 15 ml; aniline blue W.S. (1% aqueous), 1 ml; and glacial acetic acid, 4 ml. The films are washed rapidly and dehydrated in 98% ethanol. Permanent preparations may be made by mounting in Euparal. The total population of microorganisms can be determined since the depth is known and the area of the field can be determined with a stage micrometer. The bacterial cells normally present in the agar are counted in control films prepared without soil. Mycelial fragments may be counted and measured for total length under the microscope (200X). A measuring scale can be inserted in the eyepiece and the quantity of mycelium expressed as total length per gram of soil. Other methods of measuring mycelial length include the projecting of images of the fields examined and measuring the hyphal fragments with a map measuring device, or by means of camera lucida drawings of several fields and the total amount of hyphae per field determined with a map measuring device.

4. Vital staining (Strugger, 1948).

When soil is stained with the fluorescent vital dye, acridine orange, and examined under a fluorescence microscope, it is possible to observe and count living bacteria in soil in their autochthonic condition. Many of these bacteria cannot be cultured in ordinary media. They are small (0.3-1 μ) and can be observed only with high-quality microscope objectives. Using the acridine-orange technique, Strugger found that soil contained 500-10,000 million living bacterial cells per gram. His technique follows.

Ten-ml portions of acridine-orange solution are added to 1-g samples of soil in test tubes. As different soils vary in their ability to absorb the stain, preliminary trials with concentrations of acridine orange from 1:1,000-1:5,000 should be made. The suspensions are shaken and observations made relative to absorption of the stain. If the concentration is too low, the solution will be totally discolored within a few minutes and the humus ingredients of the soil will not be stained red enough. If the concentration is too high for the soil concerned, excess stain remains in the solution, hindering observation. The concentration is satisfactory when only a little excess stain remains in the solution after vigorous shaking.

Five to 10 min after shaking, the suspensions may be examined under a luminous blue light fluorescence microscope. An ordinary glass slide with coverslip or a hemacytometer slide may be employed for examination and counting. If the preparations are made correctly, the particles of soil covered with humus and the particles of humus glow in dim red fluorescence and the bacteria living on the particles appear green. To reduce the red fluorescence of the background soil particles, sodium pyrophosphate at the rate of 1% is added to the soil suspension prior to adding acridine orange (Zvyagintsev, 1964). This substance serves effectively as a quencher and living cells become more distinct. An added advantage is that a very strong light source becomes unnecessary.

B. BURIED SLIDES

Glass slides buried in soil become coated with the soil solution, somewhat like quartz grains. Vicinal microorganisms and those which spread through the soil come in contact with the slide and grow along its surface, adhering to it in the film of moisture. The organisms adhering to the glass can be stained and examined microscopically. One of the chief disadvantages of this method is that some microorganisms adhere to the glass more tenaciously than others (Zvyagintsev, 1959). Some bacteria are washed off easily, while others may resist even a strong jet of water. The number of cells adsorbed per unit surface depends on the species present, provided other conditions are equal.

1. Uncoated glass slides (Rossi and Riccardo, 1927; Cholodny, 1930).

A slit is made in the soil (in the field or in containers in the laboratory) with a sharp knife, and a clean glass slide is inserted in the slit. The soil is pressed gently to bring it in contact with the slide which is left in position for 1-3 weeks.

After soil is removed from one side, the slide is removed by gently breaking it away parallel to the soil surface. The undisturbed side is washed gently in a small stream of water to remove excess soil and the bottom side is cleaned with a cloth. It is then air-dried and fixed over a low flame. The slide is placed over a steam bath (beaker of boiling water) and stained for 5-6 min in the following stain:

Erythrosin (or Rose Bengal)	1.00 g
Aqueous phenol (5%)	100.00 ml
Calcium chloride	0.05 g

The stain should not be allowed to dry during this period. After the staining process the slide is washed, dried, and examined microscopically.

2. Vital staining (Lehner et al., 1958).

Instead of the usual fixing and staining, slides (after being removed from the soil) are dipped in a 1:750 solution of acridine orange, washed, and thus made ready for fluorescence microscopy. This combination of the buried slide and Strugger's techniques (Section A4) permits the examination of soil microorganisms in their natural structural state. The living organisms fluoresce light green against the brick-red soil particles.

3. Slides coated with agar media (Wright, 1945).

Sterilized slides are placed in sterile petri dishes and coated with malt agar. When the agar has hardened, the slides are removed aseptically and slowly dried on an electric plate at 45-50 C until the agar is reduced to paper thinness. The agar-coated surface is placed next to the soil profile in the usual manner for 3-5 days. During this time the dried malt agar absorbs enough moisture to become softened, and thus furnishes a suitable substratum for the growth of microorganisms The slides are removed from the soil, washed gently, dried, and stained for detailed examination with a microscope.

4. Pedoscopes (Gabe, 1961).

A pedoscope consists of a set of glass capillary cells set in a glass holder. Each cell has five rectangular channels, arranged in a row under a thin flat cover. A pedoscope is inserted in soil with the glass holder and may be left for several days or even months. When the pedoscope is removed from soil, microorganisms that have developed in the flat, rectangular capillaries can be viewed in a living condition with a microscope. The technique is based on the principle that natural soil is physically similar to an open capillary system. Use of pedoscopes enables study of associations of microorganisms under conditions close to those found in nature. For details of construction and use of the capillaries, see the original reference.

Aristovskaya (1962) coated the walls of the capillaries with a thin layer of an organo-mineral complex of humic acids and sesquioxides. It was found that many microorganisms, previously undescribed, would develop in the capillaries on this medium. The glass walls of the capillaries and the humus medium served as a model for soil particles. For preparation of the humic acid compounds, see Aristovskaya and Parinkina (1961).

C. BURIED NYLON GAUZE (WAID AND WOODMAN, 1957)

Some of the defects of the buried-slide method can be overcome by use of nylon gauze instead of glass slides when estimates of hyphal activity are desired. The smooth, even surface of a glass slide is quite different from the irregular, discontinuous structure of soil particles. Gauze with 12 meshes per cm permits the movement of gases and water and penetration by small roots, and presents no serious obstacle to the movement of small soil animals.

Samples of nylon gauze are buried vertically or horizontally in the soil and left for various periods of time (up to one month). Care should be taken not to touch the buried portion of the gauze or the soil particles adhering to it during its removal from the soil. All roots are carefully cut on both sides of the gauze samples, which then can be preserved in specimen tubes containing formal-acetic water.

The hyphal density within the spaces formed by the nylon gauze can be measured with a microscope in several ways. One method is to count the number of individual hyphae observed within each mesh. Data become unreliable when more than 10 hyphae occur per mesh. A statistical analysis has shown that recording the presence or absence of hyphae per mesh is an adequate and a less laborious method of counting and is reliable when less than 80% of the meshes contain hyphae.

D. THE IMPRESSION SLIDE (BROWN, 1958)

An impression-slide technique was designed to study variations in the mycelium occurring in different dune soils or soils with low organic matter content. It is useful for obtaining quantitative data on a comparative rather than on an absolute basis. Soil profiles are prepared for study by removing exposed soil from the vertical face of the profile with a sterile scalpel to prevent contamination of one horizon by another. Sterile glass slides are smeared with nitrocellulose thinned to a suitable consistency with amyl acetate. To prevent brittleness, castor oil is added to the mixture (concentration of 5% by volume) prior to smearing. The adhesive is spread with a brush over a 5 × 2.5 cm area near the center of each slide just before it is pressed for 20 sec to the soil profile. Variations in the adhesive powers of the soil constituents and in the amount of pressure applied to the slides may alter the density of the soil films. Therefore, it is important to spread the adhesive thinly and evenly and to press the slides lightly so that only the material flush with the soil surface adheres. Excess material not touching the adhesive is removed by gently tapping the slides when the preparations are dry. The soil films are stained for 1 hr in phenolic aniline blue (formula in Section A3 of this Chapter), rinsed quickly in sterile distilled water, and dried.

The dry mounts are examined with reflected and transmitted light. An 8-mm metallurgical objective, bloomed and corrected for uncovered mounts, and a compensating ocular eye-piece are used for examination. Presence or absence of mycelium is recorded in random microscopic quadrants. The quadrants can be delimited with an ocular micrometer in the eyepiece. The soil films should be examined at all depths of focus before a recording is made.

E. COUNTING SOIL ALGAE (TCHAN, 1952)

This method is based on the premise that, with a few exceptions, active algal cells contain at least one of the chlorophylls. The red fluorescence of chlorophylls illuminated with blue or ultraviolet light can be utilized for detecting and counting algae.

A soil suspension is made; the dilution depends on the nature of the soil and the prior environmental conditions. A drop of the suspension is placed on a hemacytometer slide with a depression of 0.1 mm, and a coverslip is put in position. The suspension is examined with a microscope and illuminated through a condenser (1.4 aperture) by a 6- or 12-volt lamp, the light of which passes through a solution of copper sulfate in aqueous ammonia. A yellow filter below the eyepiece absorbs blue light. Objectives of 10X or 20X are used with a 5X eyepiece.

Algae appear as red spots or lines against a black background. Soil particles are usually invisible, but sometimes they fluoresce yellow or green. Since the volume, area, and dilution factor are known, the number of algal cells per gram of soil can be calculated. For precise work the moisture content of the original soil sample is taken into account when calculations of numbers per gram are made.

F. IMMUNOFLUORESCENT-STAINING TECHNIQUE (SCHMIDT AND BANKOLE, 1962; KUMAR AND PATTON, 1964; EREN AND PRAMER, 1966)

Immunofluorescent staining may be valuable for studying the presence and distribution of specific organisms in soil or on plant roots. Briefly, the procedure consists of inoculating rabbits with the organism in question, obtaining serum from the rabbits which contains antibody protein, coupling the antibody protein to a fluorescent dye, and then treating the organism (the specific antigen) in soil, on roots, or on contact slides previously buried in soil with the antibody-dye complex. If the antibody is specific, spores and hyphae of this particular organism will fluoresce and can be observed with a fluorescence microscope. Several limitations to the method have been encountered, the most important of which is that clay particles in soil decrease the intensity of fluorescence and interfere physically with observations. Soils with more than 5% clay generally are not suitable. Details of the rather involved procedure are found in the article by Eren and Pramer (1966).

G. FLOTATION OF SPORES (LEDINGHAM AND CHINN, 1955)

A flotation method is suitable for extracting and counting number of spores of *Helminthosporium sativum* in soil and also large spores of other fungi, such as *Alternaria* and *Trichothecium*.

Water is added to a sample of screened soil to bring the soil to a moisture level of 10% by weight. A 10-g sample is mixed with 5 ml of mineral oil (Nujol) in a watch glass. The mixture is transferred to a test tube, 25 × 250 mm in size, and 50 ml of tap water are added. The tube is agitated vigorously for 4 or 5 min, then placed in a vertical position.

In about 30 min, most of the soil settles and an emulsion collects at the surface. Samples of the emulsion are transferred to microscope slides for examination. Droplets of oil and small air bubbles in the emulsion on the slide are broken down by agitation with a needle, and the mixture is spread in a film. A sample of the emulsion is obtained with a pipette which delivers drops of 1/50-ml volume. The pipettes may be drawn from glass tubing. The spores in at least ten drops are counted, and the approximate total number of spores in the emulsion (and hence spore no./g soil) are computed. A wide-field binocular with a magnification of 72X is used for observing and counting the spores. The term "flotation count" is applied to spore numbers determined with this method.

H. HYDROFLUORIC ACID EXTRACTION (NELSON AND OLSEN, 1964)

Hydrofluoric acid will dissolve extraneous siliceous material in soil and aid in freeing spores of certain fungi from mineral material, so that they may be seen in soil smears on microscope slides. Resting sporangia of *Synchytrium endo-*

bioticum were recovered efficiently from both artificially and naturally infested soil. The method was compared experimentally with flotation methods using chloroform, dibromoethane, or oil mixtures. Significantly higher counts of sporangia were obtained with the hydrofluoric acid method and counts were less variable than those obtained with flotation procedures. The method is as follows:

A 2.5-g sample of infested soil is placed in a preweighed polyethylene centrifuge tube (25 × 105mm) and half-filled with concentrated hydrofluoric acid (48-51% HF). Soil and acid are mixed at least 4 times during a 48-hr holding period in a fume hood. Then, distilled water is added cautiously to bring the liquid level to about 13 mm from the rim of the tube. The contents are mixed and centrifuged for 15 min at 2,800 rpm (1,200 G.) and the supernatant is discarded. The soil plug is thoroughly washed with distilled water and the liquid level is restored. The material is centrifuged for 11 min and the supernatant is discarded. Four more such washings are performed. The final supernatant is discarded and the weight of the washed soil is determined. A major portion of this wet extracted soil is weighed into a 50-ml beaker and diluted 10-20 fold with distilled water. To aid the suspension of soil colloids, 1-2 drops of saturated $Ca(OH)_2$ solution are weighed into the soil-water mixture. While the suspension is in motion on a magnetic stirrer a loopful is removed and weighed on a previously weighed microscope slide. A coverslip is applied and water is allowed to flow under the coverslip to disperse the sporangia. The resting sporangia in the area under the coverslip are counted at 128X magnification. From the microscopic examination of several such slide preparations, the average number of resting sporangia/g of soil is computed.

I. SOIL SECTIONING

For studying soil microorganisms and their physical relationship to each other and to the nonliving constituents of soil, thin sections of soil cut with a microtome can be examined with a microscope. Early attempts at cutting thin sections of soil resulted in shriveling to such an extent that many of the organisms became unrecognizable. Two newer methods described below have been used with success for cutting sections 7.5 μ or more in thickness without disturbing the original positions of the soil particles. The organisms are not physically distorted and can be readily observed after staining. Nematodes, fungi, bacteria, and amoebae are easily distinguished from soil particles.

1. Gelatin-embedding technique (Minderman, 1956).

A metal box (Fig. 3-1) is driven into a flat-surfaced wall of a small pit dug in the field. A space approximately 5 mm wide between the pit wall and the back side of the box is left for the gelatin solution. A 15-20% solution of gelatin at a temperature of approximately 50 C is poured into a funnel placed into the 5-mm space between the pit and the box. A thermosflask is convenient for carrying

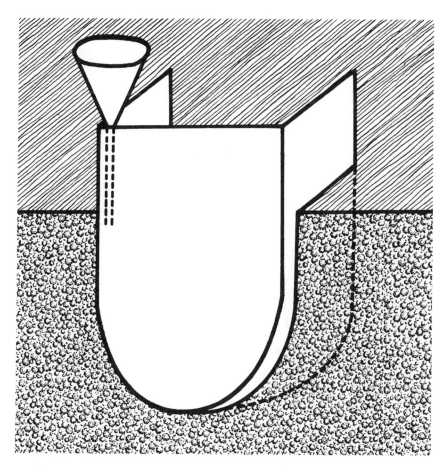

FIG. 3-1. Box driven into vertical profile for gelatin embedding soil. (Minderman, 1956. Plant and Soil 8:42-48.)

(and maintaining the temperature of) the gelatin solution in the field. More gelatin solution should be poured in after some settling, until the whole space is filled and even the litter is impregnated.

When the gelatin is sufficiently cooled and hardened, the impregnated soil plus the box is dug out. In the laboratory the box and soil are immersed in 10% formalin for at least 1 week for hardening of the gelatin. The soil is then removed from the box and areas not impregnated with gelatin are washed away.

The hardened sample is cut into appropriate sections, placed in perforated polyethylene tubes, and immersed in 50% hydrofluoric acid. The plastic vessel containing the acid and samples is stored at a temperature of 15 C or less for 1-7 days. Storage period depends on the amount and nature of sand particles in the soil. When all the sand particles are dissolved, the samples are washed with water or very dilute ammonia to remove the acid. This washing procedure is continued until the wash water has reached pH 3.0 or above.

The washed samples are again immersed in 15-20% gelatin and, when cooled, samples of appropriate size are cut. These are again immersed in formalin solution for 1-2 days to fix the gelatin. This is followed by immersion in methyl alcohol (80-90%) until the gelatin is hard enough for cutting with a microtome (about 24 hours). It is best to replace the microtome knife with a safety razor blade, since one is never certain that all sand particles have been dissolved. Sections can be cut as thin as 7.5 μ and are transferred to 4% formalin solution or directly to a slide smeared with adhesive. The sections are aligned on the slides and are firmly pressed by hand beneath several layers of filter paper. Slides thus treated are placed overnight in a desiccator partly filled with 10% formalin to insure good adhesion of the sections to the slides. Good staining results can be obtained with the quadruple staining method of Johansen (1940) followed by mounting in Canada balsam. The staining schedule is as follows: Safranin, 48 hr; methyl violet, 15 min; fast green FCF, overnight; and orange G, 30 min. It is necessary to replace the ethyl alcohol of the staining solutions with methyl alcohol in order to minimize shrinkage.

2. Polyester resin-embedding technique (Nicholas et al., 1965).

Blocks of soil, about 2.5 × 1 × 1 cm, cut from larger blocks taken from soil in the field, are plunged into liquid nitrogen to obtain a quick freeze and to prevent distortion while drying. The samples are freeze-dried and then impregnated with "Marco Resin" mixture (an unsaturated polyester manufactured by Scott Bader & Co., Ltd., London). The mixture consists of 80 ml resin, 16 ml monomer, 1 g catalyst paste, and 3 ml accelerator mixed together in that order; this forms the necessary rock-hard solid within 13 hr at room temperature. Freshly prepared resin is poured into the compartments of a polyethylene ice cube tray and the freeze-dried soil samples are carefully added. The tray is then placed in a vacuum desiccator and the soil is impregnated with resin under reduced pressure (approximately 30 min at 200 mm Hg, during which time the samples are subjected to bursts of lower pressure down to 50 mm Hg). The samples are gradually (30 min) returned to normal atmospheric pressure. The impregnated soil samples are left to harden at room temperature.

After the blocks have hardened, they are removed from the trays and cut with standard geological rock-slicing equipment. Slices, 1-mm thick, are cut and then rubbed down to any desired thickness and polished with "Carborundum" and water. A convenient size is about 50 μ thick; the actual thickness can be determined with a polarizing microscope.

The above technique has been successfully used to determine the amount of fungal hyphae in soil. Data may be taken relative to the presence or absence of hyphae per microscopic field. Also, lengths of individual hyphae can be measured and total length calculated per unit volume of soil. It is possible to discern six morphological groups of hyphae as follows: (1) short fragments of dematiaceous hyphae; (2) dematiaceous hyphae *in situ;* (3) thin hyaline, septate hyphae; (4) broad, aseptate, hyaline hyphae; (5) broad, septate, brown-stained hyphae; and (6) sparsely septate fragments of purple-black hyphae. Nicholas et al. (1965) related some of these groups to species found in soil.

CHAPTER 4
MICROORGANISMS IN THE RHIZOSPHERE

The rhizosphere may be defined as that portion of the soil which is adjacent to the root system of a plant and is influenced by the root system. The width of this zone of soil varies with the type and age of the plant, and with soil environmental conditions. The most striking influence that roots exert in the rhizosphere is the stimulation of various types of microorganisms. This stimulation, or "rhizosphere effect," results in higher numbers of microorganisms and in a distinctive microbial complex which differs from that of root-free soil. Generally, higher numbers of bacteria, actinomycetes, and fungi occur in rhizosphere soil, but fungi are least affected. One of the most characteristic rhizosphere effects is the preferential stimulation of bacteria requiring amino acids for maximum growth. Gram negative rods occur more abundantly in the root zone, but gram positive rods, cocci, spore formers, and anaerobic bacteria are not usually affected.

The chemical and physical nature of the root zone is quite different from that of soil away from the root, and the biology of this complex zone has been studied extensively. Materials excreted from roots and nutrients liberated during decomposition of sloughed root cells are utilized by root-zone microorganisms resulting in their stimulation. Various compounds such as amino acids, vitamins, sugars, and tannins, occur in root exudates. These excreted compounds have a selective effect on microorganisms within the root zone and result in the buildup of species that can utilize them as food sources. Some root exudates contain chemicals that affect certain species adversely, leading to decreased numbers of these sensitive species. Microorganisms in the rhizosphere may exert profound influences upon the plant itself by decomposition of organic matter, by affecting uptake of nutrients by the plant, by associative and antagonistic relationships, and by actual parasitism of plant roots.

The term "rhizoplane" was proposed by Clark (1949) to refer to the immediate external surface of plant roots together with any closely adhering particles of soil or debris. Rhizoplane microorganisms may be isolated from root scrapings or serial-root washings, and data thus obtained are related to a root surface basis and not to a soil basis. Since the limits of the rhizoplane and rhizosphere zones are indistinct and may vary with the plant species, plant age, and soil conditions, and since isolation techniques have not been standardized, attempts to evaluate and compare data obtained by different workers have met with little success. To emphasize this difficulty, Clark pointed out that densities of microorganisms in the rhizosphere have been expressed as numbers (a) per gram of root surface scrapings; (b) per gram of gross sample, i.e., whole roots (or macerated roots) with adhering soil; (c) per gram of adhering soil fraction of the gross root sample; and also, (d) as numbers in successive washings of the root sample, and (e) as numbers per square cm of root surface.

The methods described below are some of the most common ones that have been used with success to isolate and study rhizosphere microorganisms. In cases where agar media are used, the reader is advised to refer to the section on selection of an agar medium in Chapter 2. One should keep in mind the limitations of the particular method chosen and adjust his conclusions accordingly.

A. ISOLATION FROM THE RHIZOSPHERE

1. Dilution-plate techniques.

An estimate of total populations of bacteria, actinomycetes, and fungi can be obtained by dilution and subsequent plating of rhizosphere soil on agar. This estimate of "numbers per gram of rhizosphere soil" refers to the number of viable cells or mycelial fragments in the sample capable of growing on the agar medium used in the test. Selective agar media designed to isolate specific groups of microorganisms are listed in Chapter 2. Advantages and limitations of the dilution-plate technique also apply to isolations from the rhizosphere. After plate counts are made, numbers of microorganisms per gram of rhizosphere soil can be computed and random isolates may be transferred to tubes of media for further study. A numerical value, the R/S ratio (Katznelson, 1946), may be used to compare the population of microorganisms in the rhizosphere with the population found outside the rhizosphere. This value is obtained by dividing the number of organisms per gram in rhizosphere soil by the number per gram in nonrhizosphere soil.

Details of the dilution-plate technique are described in Chapter 2. The dilution procedure for rhizosphere soil is similar to that used for nonrhizosphere soil except for obtaining soil samples and for methods of determining weights or amounts of soil used in the dilution series..

a. **Determination of soil weight after dilutions are made (Timonin, 1940).** This technique is fairly accurate and is often used. Blocks of soil containing plant roots are cut out and gently crushed, with as little tearing of the roots as possible. The roots are lifted and gently shaken to remove superfluous soil. They are placed, along with adhering soil particles in weighed flasks, each containing 100 ml of sterile water. After thorough shaking of the suspensions, dilutions are prepared in the regular manner. Since the weight of soil in each of the dilution flasks is not known (at this time), 1-ml samples are plated from each of several dilutions. Counts or isolations from the dilution plates are made from those that contain a proper number of colonies.

To determine weight of rhizosphere soil, the roots are removed from the original dilution flask, washed, and the wash water is collected in the original flask. The water is evaporated on a water bath and the soil residue is dried to constant weight in an oven at 105-110 C. The flask containing dry soil is weighed and dilution factors are calculated, allowance being made for the amount of soil removed in preparing the dilutions.

b. **Volume displacement (Reyes and Mitchell, 1962).** Roots plus adhering soil are placed in a graduated cylinder containing 18 ml of sterile, distilled water and the cylinder is shaken vigorously. The roots are removed and the process is repeated with additional roots until the total volume of soil and water is 20 ml. Thus, the size of the rhizosphere soil sample is determined by volume displacement. The 1:10 dilution of soil (v/v) is then diluted in the usual manner.

2. **Soil-plate technique (Parkinson, 1957; Parkinson and Thomas, 1965).**

This modification of Warcup's soil-plate method for isolating fungi from the rhizosphere consists of carefully removing roots from the soil and gently shaking excess soil from them. Any aggregates of soil remaining on the roots are dislodged and discarded. The closely adhering rhizosphere soil is then collected by placing the roots in a container and shaking them vigorously. Additional rhizosphere soil may be collected by spreading out the root system on a glass plate and removing small particles of soil with tweezers or a small spatula. Small samples of rhizosphere soil (0.005-0.01 g) are transferred to petri dishes with a microspatula. The approximate amount of soil transferred to each dish should be determined by preliminary trials. The samples are broken up and distributed in the dishes, after which 10 ml of cooled melted agar medium are added to each. Before the agar solidifies the dishes are rotated to distribute the soil in the medium. Fungi that grow from the soil particles after appropriate incubation are isolated and identified.

Two types of data can be obtained by using this technique: (1) Total rhizosphere isolations. Whole root systems of a number of plants are shaken and scraped carefully and a series of soil plates is prepared with the soil obtained. A number of replicate series of soil plates can be made. (2) Zonal rhizosphere isolations. Soil particles are taken from defined parts of the root system; hence, a group of soil samples from one root or from defined positions on the root system may be obtained and a series of soil plates is prepared from each sample. Replicate series of plates can be prepared by sampling from several plants.

3. Soil-washing technique (Parkinson and Thomas, 1965).

Plants are removed from the soil and shaken to loosen excess soil. The roots are then cut off at the crown and placed in screw-capped bottles containing sterile water. These bottles are shaken vigorously a few minutes until much of the closely adhering rhizosphere soil has been dislodged from the system. The roots are removed from the bottles and the suspension of rhizosphere soil is introduced into a soil-washing box via a filter funnel connected to a perspex (plexiglas) tube entering the top of the washing box. The soil-washing procedure is then followed as outlined in Chapter 2. This procedure usually provides for increased frequency of isolation of sterile hyphae, *Fusarium* spp., and members of the Mucorales from the rhizosphere. Other abundantly sporulating fungi are usually isolated with decreased frequency.

B. ISOLATION FROM THE RHIZOPLANE

1. Dilution procedures.

Louw and Webley (1959) suggested that, for isolation of microorganisms from the rhizoplane, roots in the dilution flasks containing rhizosphere soil (see Section A1 above) be removed and immersed in other flasks containing weighed amounts of sterile water plus glass beads of 2-mm diam. The flasks are reweighed and shaken for 20 min. Dilutions are prepared and dilution plates are made as described above with both rhizosphere and rhizoplane soil. The number of microorganisms obtained from rhizoplane soil is recorded as the number per gram of wet weight of root tissue.

Cook and Lochhead (1959) described a root-maceration technique. Roots are removed from dilution flasks containing rhizosphere soil and washed in three changes of sterile water. They are weighed and resuspended in sterile water and ground for 3 min in a Waring Blendor. Dilutions are prepared from the blended material and from the original rhizosphere soil suspension as described above. The number of microorganisms obtained from the rhizoplane is recorded as the number per gram of macerated root tissue.

2. Serial-root washings (Harley and Waid, 1955).

For isolating fungi intimately associated with root surfaces, root pieces are subjected to a series of 30 or more washings in sterile, distilled water. Completeness of removal of fungal spores and mycelial fragments is determined by plating samples of any number of washings on agar plates. After washing, excess surface moisture is removed from the roots by blotting with sterile filter paper. The roots are then cut into segments, 2-mm long, and placed on agar media in petri dishes. Fungi developing from the segments are isolated in pure culture and identified. This method permits the isolation of slowly growing fungi that are present as mycelia on root surfaces which are normally prevented from appearing on dilution plates by competition of rapidly growing fungi.

3. Root maceration (Stover and Waite, 1953).

Although originally designed to isolate *Fusarium* from root tissue, this technique has been used successfully for the study of saprophytic fungal colonization of plant roots. The method consists of washing roots in sterile water, adding these washed roots to an additional volume of sterile water, and macerating them in a Waring Blendor for about 2 min. Samples of the blended material are spread on the surface of agar in petri dishes. The amount of tissue to be used for blending and the amount of blended material transferred to agar plates must be determined previously by trial. Satisfactory results for isolation of fungi have been obtained with Martin's peptone-dextrose agar containing rose bengal and streptomycin, and with Czapek-Dox agar plus yeast extract (pH 5.0).

A modification of the technique consists of washing the roots serially 12 times before maceration in a Waring Blendor (Singh, 1965). Samples of the blended material are placed in petri dishes prior to adding cooled, melted agar. Before the agar has hardened, the dishes are rotated to disperse root material. This procedure combines advantages of both the serial-washing and root-maceration techniques and is especially useful when using plant roots that consist of many small diameter "hair roots." Such roots may still be heavily contaminated after serial washing.

4. Root dissection (Waid, 1956).

Roots are gently washed free from soil and then washed through 10 changes of sterile water. Small pieces of root, about 2 mm in length, are cut off and placed in sterile water in petri dishes. Under a dissecting microscope roots are dissected into two parts, outer cortex, and stele with some inner cortex. Each part is placed in separate dishes, and each root piece may be dissected into smaller fragments. Fragments and intact root pieces are then placed on agar medium (Czapek-Dox, acidified) in petri dishes and incubated for 3 weeks at 25 C. The spatial relationships of different fungal species, both saprophytic and parasitic, on and within root tissue may be studied.

C. SPECIAL DEVICES FOR COLLECTING RHIZOSPHERE SAMPLES

1. Rhizosphere sampler (Papavizas and Davey, 1961).

This sampler permits collection of rhizosphere soil at various distances from the root surface. Densities and kinds of microorganisms that occur at these locations can be determined. The sample collector (Fig. 4-1) may be used with plants that produce a straight primary root and secondary roots with predictable direction of growth and orientation in soil (such as *Lupinus angustifolium*, blue lupine). In blue lupine sets of secondary roots grow from the primary root following the same directions as the two cotyledons which rise from the soil with the primary leaves, whereas the third set of secondary roots develops between the other two (Fig. 4-2). It is not feasible to use the sample collector with plants that have fibrous root systems.

The sampler is designed for taking rhizosphere samples outward from the rhizoplane of blue lupine seedlings in 3-mm increments. It is fabricated in 2 parts, the sample collector and the sample ejector (Fig. 4-1). The sample collector consists of 7 thin-walled steel tubes, 50-mm long with an inside diam of 3 mm, soldered adjacent to each other in a plane with an overall width of 22 mm. The sample ejector is made by setting 7 steel rods, 51 mm long with an outside diam of 2.75 mm, in a 6 X 6 X 40-mm brass bar in positions to match the tubes of the sample collector.

The top 2 cm of soil above the seedling are removed and the sterile sample collector is inserted vertically into the soil immediately adjacent and parallel to the primary root of the seedling (Fig. 4-2). The sample collector is removed from the soil and the sterile sample ejector is used to deliver soil cores of desired lengths into sterile water blanks. Sample weights can be adequately controlled by collecting 1 mm of soil-core length for each 10 mg of soil sample desired. Actual oven-dry sample weights can be determined and data corrected accordingly. Soil in the water blanks can be diluted and microorganisms isolated in the regular manner.

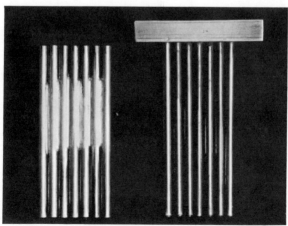

FIG. 4-1. The rhizosphere microsampler. Left, sample collector. Right, sample ejector. (Papavizas and Davey, 1961. Plant and Soil 14:215-236.)

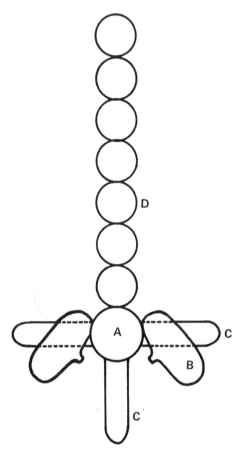

FIG. 4-2. Diagram illustrating the relation between lupine seedling and the microsampler (top view): (A) Stem or primary root; (B) Cotyledon; (C) Secondary root; (D) Microsampler (Papavizas and Davey, 1961. Plant and Soil 14:215-236.)

2. Soil box (Parkinson, 1957).

This device is an adaptation of Chester's immersion tube and Thornton's immersion plate for the isolation of fungi that produce actively growing hyphae in the rhizosphere. Wooden boxes that measure 60 × 30 × 22 cm are constructed with one face composed of closely fitting, but separate and removable, perspex (plexiglas) segments. Adequate drainage holes are made in the base, and the inside of the box is waxed thoroughly. A duplicate set of perspex segments is made for each box and the segments are labeled so that their exact positions on the face of the box are known. In this way close fitting of the segments is maintained.

The boxes are filled with soil, and seeds are planted. The boxes are turned so that the perspex face forms the base of the box. Thus, developing roots grow onto the perspex segments. When roots are observed in contact with this face, the box is returned to its normal position and the roots grow down the perspex face.

If isolation of fungi from the rhizosphere of several roots at a particular depth in the soil is desired, the perspex segment at the required depth is removed and replaced by one which has been perforated. The perforations are rebated, corresponding to the points on the various roots from which isolation of fungi is desired. Small tubes, 5 cm long with an inside diameter of 0.5 cm, are plugged at one end with cotton and half-filled with 2% water agar (pH 6.0). These tubes are placed into the perforations. The rebating employed in the preparation of the perforations is designed to minimize the possibility of contact between the agar and soil particles. The perspex face is protected by a metal shield, blackened on the inside, to eliminate any anomalous effects that might be caused by illumination of the root system.

At desired intervals (the exact time being determined by preliminary experiments) the tubes are removed from the perspex segments and are replaced by their corresponding unperforated segments. In the laboratory the cotton plug in each tube is removed, and the agar is pushed out into petri dishes with a sterile swab. In each dish the agar cylinder is cut into 4 pieces, each of which is placed in a separate petri dish where it is cut into 4 segments. Nutrient agar is added, and fungi that develop are isolated and identified.

D. STUDY OF ORGANISMS IN THE RHIZOSPHERE WITHOUT ISOLATION IN PURE CULTURE

1. Contact slide (Starkey, 1938).

An adaptation of the Rossi-Cholodny buried-slide method was designed to study microorganisms associated with roots. The method is particularly suitable for studying the development of microorganisms in response to root growth, localization of microbial individuals and groups, and the sequences of morphological types.

Microscope slides are inserted in soil in either a vertical or horizontal position. The depth at which the slides are placed will vary with the kind of plant involved and the information desired. Seeds or seedlings are planted above the slides. A portion of the experimental area is left free of plants to serve as controls. After appropriate root development the slides are taken from the soil and one side is cleaned. The larger soil particles are removed from the other side, and the slides are dried and stained. Generally, phenolic rose bengal stain is used (Chapter 3). The slides then are examined microscopically.

2. Impression slide (Parkinson, 1957).

The impression-slide technique developed by Brown (1958) was adapted by Parkinson to study mycelial development of fungi in the rhizosphere. Plants under study are removed from the soil, and excess soil is shaken from the roots so that only soil closely adhering to the roots remains. Individual roots are placed on microscope slides previously coated with a thin layer of nitrocellulose dissolved in amyl acetate. This layer provides an adhesive surface to which the rhizosphere soil sticks when the root is removed. Slides prepared in this manner are left to dry, then are stained in phenolic aniline blue for 1 hr, washed with water, and blotted dry. The material is mounted in lactophenol under a coverslip and examined under a microscope with a 10X eyepiece and 2/3 objective. The occurrence and extensiveness of mycelium in the rhizosphere can be compared with that on impression slides prepared with soil outside the rhizosphere.

3. Direct microscopic examination of roots.

Fungi can sometimes be observed on root sections previously washed, dehydrated, stained, impregnated with paraffin, and mounted with balsam. This method requires considerable time and results are not always satisfactory. The following two methods have been used with success.

    a. **Mounts in lactophenol (Parkinson and Clarke, 1961).** Roots to be examined are washed in 10 changes of sterile water. They are cut into small segments 2 mm long and stored in formalinacetic-alcohol preservative until ready for use. Prior to direct observation the roots are transferred to lactophenol for several hr, then stained in 0.5% cotton blue in warm lactophenol for 2 min. After the excess stain is removed in lactophenol, the roots are mounted on microscope slides in lactophenol. Each segment is examined for the presence of hyphae and, the type (dark or hyaline) is recorded.

    b. **Nitrocellulose replicates (Moreau and Effenterre, 1961).** Roots to be examined are washed in several changes of sterile water. They are immersed carefully in a varnish composed of nitrocellulose dissolved in a mixture of ketones and chlorinated solvents with the addition of a plastifier. The roots are left in this varnish for 12 hr, then removed gently and suspended in air. When the varnish has hardened on the root surfaces (about 1 hr) the thin films are placed under a dissecting microscope and carefully removed from the roots. The films are mounted on microscope slides and examined. These films are replicates that are exact negative microcasts of the root surface. Images of fungal mycelia, root cells, and soil particles may be observed under the microscope.

4. Root-observation box.

A soil box (Fig. 4-3) made entirely of plexiglas is designed for growth of seedlings. The box (76 × 57 × 102 mm deep) is constructed of plexiglas segments joined together tightly with plexiglas cement. A small hole is bored in the bottom for drainage. One side of the box consists of 4 glass microscope slides, held to the plexiglas with a small drop of balsam or other cementing materials, or with rubber bands encircling the box. After seeds are planted, the boxes are tilted so that the roots grow against and adjacent to the microscope slides. Fungal mycelium adjacent to the roots may be observed directly under the microscope using a strong oblique light source. When desired, the slides may be removed, position of the roots marked on the back of the slide, and handled as contact slides, i.e., washed, stained, and examined under the microscope (see Chapter 3). For isolation of fungi from the rhizosphere the slides are replaced with plexiglas segments cut to the same size as the microscope slides. These segments are perforated and small tubes containing agar are attached as described above. Parkinson's original procedure (Section C2) for isolating these fungi in pure culture is followed.

FIG. 4-3. Modified soil box, with microscope slides attached to one side.

CHAPTER 5
ISOLATION OF PLANT PATHOGENS FROM SOIL

An indirect method for obtaining plant pathogens from soil consists of growing the host plant in the infested soil and isolating the pathogen from the infected tissue of the host. According to the type of disease, pathogen, or host with which one is working, various conditions may be set up that increase the likelihood of obtaining adequate disease development. Manipulation of soil temperature, moisture, or pH may influence disease severity and affect isolation of the desired pathogen.

The host plant is not generally utilized with isolation methods described in this chapter, but testing the isolated organisms for pathogenicity is often necessary to separate morphologically similar strains. Ordinarily, plant pathogens are not isolated on dilution or soil plates as outlined in Chapter 2. Modifications of the procedure or changes in constituents of the isolation media have been devised to limit the growth of saprophytes. In some cases, a special pretreatment of the soil before dilutions are made is necessary to ensure isolation of the desired organism. Immersion tubes can be used for selective isolation of fast-growing fungi such as *Rhizoctonia, Fusarium,* or *Pythium*. Some plant pathogens can be isolated from soil by using plant tissues as selective baits. The plant tissue may be buried in soil or soil is placed in or on the tissue substrate. Certain Phycomycetes that produce motile spores may be isolated by placing the bait in water containing soil, but avoiding direct contact with the soil. Sclerotia may be isolated by passing soil through a series of screens of different size openings. Screening soil has also been useful for collection of soil-particle fractions, since some fungi normally occur only in soil fractions of a certain size. These fractions are subsequently processed by dilution or soil plates for isolating the organisms. A flotation method has been used successfully for isolating large-spored fungi, such as *Helminthosporium,* from soil.

The descriptions of methods that follow are selected ones which we feel can be used with success. It is not our intention to recommend one method over another, except where adequate experimental comparisons are made in the literature.

A. BACTERIA

1. *Agrobacterium tumefaciens* (Schroth et al., 1965).

A selective medium is used which contains a number of compounds and antibiotics that inhibit the growth of various fungi and bacteria but do not adversely affect the growth of *A. tumefaciens.* See Chapter 16 for ingredients. Water is added to infested soil, generally to a dilution of 1 g of soil in 10 g of water. The suspension is shaken thoroughly and left to settle for 1 min. One-tenth ml of the supernatant is pipetted onto the medium in petri dishes and spread evenly over the surface with an L-shaped glass rod. *Agrobacterium* colonies usually are visible after 2 days of incubation at 28 C, and reach a diameter of 2-4 mm after 4 days. The colonies are smooth, glistening, translucent, convex, circular with an entire margin, and yellow to salmon yellow in color. A distinctive granular appearance can be observed with a dissecting microscope and transmitted light. According to the authors, the accuracy of differentiating *Agrobacterium* colonies from other bacteria ranges from 95 to 100%.

2. *Agrobacterium rhizogenes* (Ark and Thompson, 1961).

Carrot, turnip, or beet roots are washed in water plus a suitable detergent to free them of soil. The roots are rinsed in several changes of distilled water, then partially dried, sliced with a sterile knife, and placed in petri dishes. A suspension of soil is smeared on the cut surfaces and the dishes are incubated at 25 C. In 7-14 days the bacterium, if present, causes roots to develop on the inoculated root surfaces.

3. Soft-rot bacteria (Kerr, 1953).

Freshly dug potato tubers, free from surface wounds, are surface-sterilized and washed with several changes of sterile water. The tubers are placed in a soil suspension (5 g soil in 250 ml sterile water) for 5 min, and incubated for 48 hr in moist chambers at 26 C. Soft-rot bacteria can infect the tubers through lenticels. To isolate the bacteria, a visibly infected tuber is washed with tap water and cut lengthwise with a sterile scalpel. A cut is then made in the healthy tissue of the exposed surface so that the half-tuber can be split across to expose a rotted area. This procedure is used to avoid contamination with the scalpel. A small piece of rotted tissue is transferred to sterile water in a test tube and left for 30 min. The bacterial suspension thus obtained is streaked onto nutrient agar and the bacteria from gram-negative colonies are isolated. Each isolate should be inoculated into freshly cut sterile slices of potato tuber to determine its pathogenicity.

B. *CYLINDROCLADIUM SCOPARIUM*

The following method proposed by Thies and Patton (1970) is similar to one described by Morrison and French (1969) for isolating *C. floridanum* from soil. The procedure consists of wet-sieving of soil to concentrate the microsclerotia, which are then germinated on a selective medium. A 400-g sample of naturally infested soil is put into a Waring Blendor, and the jar is nearly filled with water. The blender is run at low speed for 2 min and the suspension is left to settle for 15 sec. The supernatant is decanted onto nested 100- and 200-mesh screens (pore size 149 and 74 μ, resp.). Water is added to the residue in the blender jar, and the residue is stirred into suspension. After a settling period of 15 sec, the supernatant is decanted onto the screens. This washing and decanting procedure is repeated until the supernatant is clear, usually 6 to 9 times. The residue on the 100-mesh screen is washed and discarded. The material collected on the 200-mesh screen is washed thoroughly on the screen, then collected in a 400-ml beaker. The beaker is filled with dilute water agar (0.3% agar) and the suspension is stirred for 10 min with a food mixer. Fifteen ml of the suspension are added to 200 ml of melted and cooled agar medium. The medium is made by adding lactic acid (enough to produce pH 3.5) and 1,000 ppm of Union Carbide Corporation's TERGITOL Nonionic NPX to melted and partially cooled Czapek's medium containing 2.5% agar. The medium containing the suspension is dispensed into petri dishes.

After 10 days of incubation, the plates are examined for colonies of *C. scoparium*. This fungus is readily identified by the numerous microsclerotia that form in the agar and impart a characteristic dark reddish brown color to the colony. Estimates of the total number of viable propagules of the fungus in the original soil sample can be made.

C. *FOMES ANNOSUS*

A selective medium is used to isolate *F. annosus* from soil (Kuhlman, 1966). Soil is diluted with water; the final dilution should be determined by trial. Samples of the final dilution are plated on the selective medium (Chapter 16). Colonies of *Trichoderma* are frequently encountered on the selective medium, but they ordinarily do not prevent enumeration of the *F. annosus* colonies.

D. *FUSARIUM*

1. Soil dilution (Nash and Snyder, 1962).

This method was devised to count and isolate propagules of *Fusarium solani* f. *phaseoli* in field soil. Soil dilutions are prepared in a 0.1% agar suspension instead of plain water. The medium, PCNB agar, contains streptomycin and PCNB which inhibits spreading colonies of bacteria, yeasts, and other fungi. The ingredients of this medium and an improved PCNB medium formulated by Papavizas (1967) are found in Chapter 16. Best results are obtained by leaving the poured plates in a dark, cool place to dry for 3-5 days before they are inoculated with the soil suspension. One-ml samples of the final soil dilution (0.1% agar suspension) are pipetted to the surface of the agar in each dish, and the suspension is spread by rotating and tilting the dishes. Diffuse daylight for the first few days of incubation is necessary for profuse sporulation of some isolates of *Fusarium*. Colonies of *Fusarium* can be identified on these plates after 5-7 days of incubation.

For isolation of *Fusarium roseum* f. *culmorum* and f. *gibbosum*, phytoactin (L-318 formulation) is added to the medium after autoclaving to a final concentration of 10 ppm (Stoner and Cook, 1967). This antibiotic eliminates about 50% of the colonies of *Fusarium* other than f. *culmorum* and f. *gibbosum*. Growth of colonies is slowed but numbers obtained are not affected.

2. **Soil washing (Watson, 1960).**

One-g samples of soil are placed in 500-ml Erlenmeyer flasks, each containing about 200 ml of water. The soil is agitated thoroughly and left to settle with the flasks on a 45° angle for 1 min, and the water is then decanted. Water (200 ml) is again added to the soil residue, and the washing procedure is repeated. This procedure may be repeated 16 or more times. The soil residue which remains following the washing procedure is dispersed in water for a few sec in a Waring Blendor. The soil-dilution procedure (see Chapter 2) is then followed to isolate fungi. Fungi other than *Fusarium* will be obtained on these dilution plates, but the ratio of *Fusarium* to other fungi is increased considerably.

3. **Fusarium-culture filtrate method.**

F. solani f. *phaseoli* or *F. oxysporum* f. *conglutinans*, when grown on potato-dextrose broth, modifies the broth in such a way that solid media prepared from culture filtrates are entirely unsuitable for the growth of many fungi, and the growth of many other fungi is greatly reduced (Parmeter and Hood, 1961). This inhibition of competing fungi on soil-dilution plates prepared with filtrate media enables consistent estimation of the populations of *F. solani* f. *phaseoli* in soil.

Fifteen-day-old liquid cultures (potato-dextrose broth) of *F. solani* f. *phaseoli* are filtered and sterilized through a Seitz filter. Agar is added to this filtrate so that a 2% agar medium is obtained. It is then autoclaved for 15 min at 18 lb pressure. Streptomycin is added to the medium at 300 ppm prior to pouring into petri dishes. Soil suspensions are prepared by adding a weighed sample of soil to a known volume of 0.1% water agar. One ml of suspension is transferred to the agar surface in each dish and distributed by gentle agitation. Dishes thus prepared are examined for colonies of *Fusarium* after 5 days at room temperature.

E. *GEOTRICHUM CANDIDUM*

This fungus, which causes a sour rot of citrus fruit, can be isolated with a fragmentation method (Butler and Eckert, 1962). Two-4 g samples of soil suspected of being infested with *Geotrichum,* are placed in 40- to 50-ml vials with caps fitted with a rubber washer. The soil in each vial is covered with about 20 ml of potato-dextrose broth containing 100 µg/ml of novobiocin. The caps are screwed on tightly and the vials shaken on a mechanical shaker for 48 hr at 20-24 C, or for 24 hr at 27-30 C. The vials are then shaken vigorously by hand to suspend the contents, and about 0.3-0.5 ml from each vial is poured onto the surface of a potato-dextrose-novobiocin agar. The liquid is distributed over the surface of the agar and incubated at 27-30 C for 24 hr.

Generally, a dense mycelial mat of *G. candidum* develops on plates from soils containing the fungus. It can then be isolated to pure culture. The effectiveness of this method is attributed to the nature of the fungus to multiply by fragmentation in liquid shake-culture, and it is relatively insensitive to high concentrations of CO_2 which probably occur in the vials during shaking. In some cases, species of *Mucor* tend to overgrow the cultures. Growth of *Mucor* can be minimized by making sure that the pH of the potato-dextrose broth is 5.7-6.1, and not extending the shaking time over the recommended period.

F. *HELMINTHOSPORIUM SATIVUM*

The flotation method designed by Ledingham and Chinn (1955) is described in Chapter 3. The technique is modified as follows (Chinn et al., 1960). About 25 ml of the emulsion containing *H. sativum* is pipetted into a 30 X 70-mm screw-cap vial and warmed to 45 C. Potato-dextrose agar containing 1% molasses is melted and cooled to 45 C, and 8 ml are added to the emulsion. The mixture is shaken vigorously and a clean microscope slide is dipped momentarily into the agar susupension. As soon as the agar solidifies on the slide, it is placed in a staining dish fitted with moist filter paper. A further precaution against drying of the agar on the slide can be taken by placing the dish in a second closed container. The slide is incubated at 24 C for 15-20 hr. Agar on one side of the slide is wiped off and a dilute solution of cotton blue is sprayed on the opposite side with an atomizer. The suspension is examined under the microscope, and germinated and nongerminated spores of *H. sativum* are recorded.

G. *PHYMATOTRICHUM OMNIVORUM*

Sclerotia of this fungus may be recovered from soil with a sieving technique (King and Hope, 1932; Rogers, 1936; Streets, 1937). Soil samples are removed from definite zones at various depths by means of a trowel or post-hole auger. Each sample is placed in the top of a set of two sieves. The upper sieve has openings of 2 mm and the lower, 841 μ. The finer soil particles are forced through the sieves by water under pressure from a hose nozzle; the larger particles and root fragments are left on the top sieve. Sclerotia, with coarse sand and other debris, are caught on the bottom screen. They can be picked out easily with forceps.

To float sclerotia away from the debris on the fine screen a sugar solution with a specific gravity of 1.15-1.25 is stirred into the mixture (Rogers, 1936). The sugar solution does not affect the viability of the sclerotia. Heavy impurities can be separated from the sclerotia by immersing the sample in a saturated solution of photographers' hypo (sodium thiosulfate). The floating fraction is caught on a sieve, rinsed, and placed in a porcelain evaporating dish and the organic debris is washed away with a carefully controlled jet of water (Streets, 1937).

H. *PHYTOPHTHORA*

1. Soil dilution.

   a. ***Phytophthora megasperma* var. *sojae*** (Haas, 1964). Rhizosphere soil from diseased plants is collected and screened (pore size, 2 mm). The soil is diluted in 0.1% water agar to a final dilution of 1:1,000. Samples (0.5 ml) are placed in petri dishes and agar medium cooled to 42 C is poured; the dishes are rotated gently to disperse soil particles. The agar medium used is corn meal agar supplemented with pimaricin (2 ppm), penicillin "G" (80 units/ml), polymixin B (370 units/ml), and pentachloronitrobenzene (10 ppm). The pH is adjusted to 4.6 with lactic acid. The concentration of pimaricin is critical; direct isolation from soil is prevented at concentrations of pimaricin higher than 2 ppm even though *Phytophthora*, transferred from pure culture, grows at 100 ppm. Acceptable restriction of other fungi is not attained with pimaricin at less than 2 ppm.

   b. ***Phytophthora parasitica* and *P. citrophthora*** (Ocana and Tsao, 1965; 1966). These fungi can be isolated from rhizosphere soil of diseased citrus seedlings with the method of Haas described above. The Difco corn meal agar used is supplemented with pimaricin, Vancomycin, and pentachloronitrobenzene (PCNB) at 10, 200, and 100 ppm, respectively. At these concentrations, Vancomycin effectively suppresses soil bacteria and actinomycetes; PCNB further enhances the control of soil fungi.

   c. ***Phytophthora cinnamomi*** (Hendrix and Kuhlman, 1965). Soil from an infested area is diluted in 0.5% water agar. The final dilution to be used should be determined by trial, but generally dilutions of soil in water of 1:2 to 1:10 are satisfactory. One ml of the diluted soil is spread on the surface of Modified

Kerr's agar medium (Chapter 16) and incubated at 20 C for 72-96 hr. Soil is removed from the agar surface under flowing tap water. Typical *Phytophthora* colonies are counted and transferred to fresh Kerr's agar medium. *Pythium* spp. may also be isolated on this medium (Campbell and Hendrix, 1967).

    d. **Phytophthora parasitica var. nicotianae (Flowers and Hendrix, 1969).** One ml of a 1:50 soil suspension (0.5 g soil/25 ml 0.5% agar) is spread over the surface of a selective medium containing gallic acid in a petri dish. Ingredients and method of preparation of the medium are outlined in Chapter 16. Dishes thus prepared are incubated for 36 hr in the dark at 24 C. The soil is washed off the agar with slowly flowing tap water. Both *Pythium* and *Phytophthora* species develop on this medium, but colonies of *Pythium* are larger and less compact than *Phytophthora*. The dense, extremely branched, knotty appearance of *Phytophthora* colonies readily distinguishes them from *Pythium* colonies when viewed under low power with a microscope. Reproductive structures necessary for identification can be obtained by transferring bits of the colonies to hemp seed agar.

2. **Soil sieving (McCain, Holtzmann, and Trujillo, 1967).**

    Chlamydospores of *Phytophthora cinnamomi* range in diameter from 40-135 μ, and may be concentrated on sieves prior to isolation on a selective medium. Fifty g of soil previously passed through a 4-mm screen are placed in 500 ml of water in a Waring Blendor and stirred for 1 min at slow blender speed (3,000 rpm). The supernatant is poured immediately through nested sieves with mesh sizes of 149, 61, 44, and 38 μ. Sediment in the blender is restirred twice in 200 ml of water for 20 sec at the slow blender speed and the liquids and solids are poured through the nested sieves. The residue on the 149-μ sieve is washed with a spray of water. The residue on the 61-, 44-, and 38-μ sieves is collected in separate containers by holding each screen at an 80° angle and carefully backwashing on the bottom side of the screen with a stream of water and then removing the residue from the sieve with a stream of water directed onto the inner surface. Residue from the 38-μ screen may be rescreened. The total volume of residue collected on each screen is poured onto a solidified medium in petri dishes at the rate of 10 ml/plate. The solidified medium contains 4% agar, 1% V-8 vegetable juice, 50 ppm nystatin (Mycostatin), 100 ppm Vancomycin, and 10 ppm PCNB (pentachloronitrobenzene). The plates are incubated at 20 C for 12 hr and then washed under a slow stream of running tap water to remove materials other than germinated spores which remain attached by hyphae to the medium. The plates are examined after 24 hr and the numbers of colonies of *P. cinnamomi* are counted.

3. **Baiting techniques.**

    a. **Apples (Campbell, 1949).** This is an adaptation of a technique originally designed for isolation of *Phytophthora* from infected tissue. Holes are bored in apples, almost through, and filled with soil to within 1 cm of the top. The remaining space is filled with distilled water. After the soil is saturated the openings are sealed with cellulose tape, and the apples are incubated at room temperature for 5-10 days. *P. cinnamomi,* if present, causes a firm dry rot from which it can be isolated and transferred to tubes of corn meal agar.

    b.   **Avocados (Zentmyer et al., 1960).** Fuerte avocado fruit can be used to trap *P. cinnamomi.* Firm fruit are half-embedded in soil in containers, and the soil is saturated with water. The containers are incubated at 25-27 C for 3-6 days. Infections appear as brown, firm, circular spots at the water line.

    c.   **Lemons (Klotz and DeWolfe, 1958).** Soil is placed in containers and water is added to saturation. Clean unsprayed lemons are placed on the soil surface. Incubation at 20 C permits growth of many species of *Phytophora.*

    d.   **Pineapple crowns (Anderson, 1951).** One to 2 tablespoons of soil are placed in quart Mason jars filled with tap water. Rooted pineapple crowns are placed in the top of the jars so that roots are in the water but do not touch the soil in the bottom. The containers are incubated at room temperature below 26 C. After about 4 days, water-soaked lesions begin to appear near the root tips. The lesions are removed, placed in a petri dish with water and examined under the microscope. Presence of *P. cinnamomi* is indicated by typical nonpapillate sporangia borne singly on long sporangiophores, accompanied by proliferated sporangia with one to several successive sporangial walls telescoped within the original or outer wall. The fungus can be transferred to pure culture if desired.

    e.   **Potato slices (Zan, 1962).** This method was designed to isolate *P. infestans* from soil. Potato tuber slices, 0.5 cm thick, are placed in moist chambers. To each slice is added 0.5 ml of soil with sufficient water to allow even spreading and to provide uniform infection conditions. The slices are incubated in the moist chambers for 5-6 days at 20-22 C and are then examined for the presence of sporangia.

I. *PYTHIUM*

1. Soil plate (Singh and Mitchell, 1961).

The procedure for Warcup's soil-plate method is followed (see Chapter 2), but with a selective medium. The basal medium used is Martin's peptone-dextrose-rose bengal agar. For bacterial inhibition, Agrimycin (15% streptomycin, 1.5% oxytetracycline, and 83.5% inert ingredients) is added to the medium so that the final concentration is 100 mg/liter. For inhibition of fungi other than *Pythium,* pimaricin is added to the basal medium to yield a concentration of 20 mg/liter. Agrimycin and pimaricin are added as sterile solutions to the medium after it has been autoclaved. Soil plates thus prepared are incubated at 25-28 C for 1-3 days. Colonies of *Pythium* are tentatively identified with the aid of a dissecting microscope and transferred to pure culture.

2. Soil-particle plates (Schmitthenner, 1962).

A modification of Reischer's medium (Chapter 16) is used with this method. Soil samples are sieved and individual particles of 1-7 mg each are selected. Particles that weigh 1 mg or more are normally retained on a sieve with openings of 841 μ. Four particles per petri dish are placed on the surface of the agar medium. The entire agar disc is inverted and the convex rim of agar at the edge of the dish is removed so that the agar surface will adhere to the bottom of the dish and seal the soil particles in small pockets beneath the agar. The dishes are incubated at 25 C for 48 hr and examined with a microscope at 200X magnification. Most fungi other than *Pythium* are inhibited, and characteristics of different species of *Pythium* are distinct enough for identification.

3. Baiting techniques.

 a. **Pineapple crowns (Klemmer and Nakano, 1962).** This bait technique is a modification of Anderson's method for isolation of *Phytophthora cinnamomi*. Soil is placed in suitable containers filled with tap water. After the soil has settled to the bottom, pineapple crowns previously rooted in tap water are partially immersed in the water so that root tips are 5 cm or more above the settled soil sample. Alternately, young heart leaves or pineapple crowns or slips are strung over the top of the containers so that the white basal tissue of each leaf is immersed in water. Both bait systems are incubated at 20-25 C. With either system, zoospores of pythiaceous fungi swim upwards and infect the plant tissues. After infection, the tissue is transferred to corn meal agar containing 100 ppm of pimaricin.

 b. **Green apples (Boothroyd, 1967).** Soil is placed in suitable containers to about 3 cm in depth. The blossom ends of green apples are punctured several times with a dissecting needle, and pressed into the soil. Water should be added so that a moist-chamber effect is obtained when covers are placed on the containers. After incubation at 20-24 C for 2-3 days, the apples are examined for decay, and the infected tissue is transferred to suitable agar media.

 c. **Sweet corn seed (Goth et al., 1967).** Seeds are soaked for 16 hr in distilled water and autoclaved for 30 min at 121 C. Soil is placed in glass containers to a depth of 2 cm and brought to 100% of its moisture-holding capacity with distilled water. Autoclaved seeds are placed in the soil approximately 1 cm deep and 2 cm apart. After incubation at 25 C for 4 hr, the seeds are washed with distilled water and the number of infested ones determined by plating on Schmitthenner's modification of Reischer's agar medium (Chapter 16). Hyphal tips of emerging fungi are transferred to corn meal agar and isolates of *Pythium* obtained are tested for pathogenicity on seedlings of the test plant. When sterile quartz sand is used as a diluent for the test soil, a dilution end point may be reached and the population of *Pythium* in the test soil (expressed as infective centers) is proportional to the number of seeds colonized.

d. **Potato cubes (Hine and Luna, 1963).** In this baiting technique selective antibiotics are utilized to inhibit other fungi and bacteria. Freshly diced potato cubes of approximately 3 mm are soaked in an aqueous solution of 100 ppm streptomycin sulfate and 100 ppm of pimaricin for 1 hr. Soil from the field is placed in petri dishes and moistened to field capacity. Cubes are placed in the soil and incubated for 7-12 hr at 30 C. The cubes are removed, rinsed thoroughly in tap water, and placed on water agar supplemented with 100 ppm streptomycin and 100 ppm pimaricin. Hyphal tips of *Pythium* spp. originating from the cubes are cut and transferred to pure culture. Excellent recovery of *Pythium aphanidermatum* from either naturally or artificially infested soils was obtained with this technique. The authors found that certain isolates of other species of *Pythium* were inhibited by streptomycin, and these could not be recovered on agar supplemented with the antibiotic.

J. *RHIZOCTONIA SOLANI*

Davey and Papavizas (1962) compared four methods for isolating *Rhizoctonia solani* from soil. Each method had certain distinct advantages and characteristics which would suggest its use under particular circumstances. The infected-host method was best for studies on the range of clones of the fungus pathogenic to certain plants. A limited number of clones of the fungus was recovered with the immersion-tube method. The debris-particle method in conjunction with the buckwheat-colonization method gave the most complete information, since the former reflected the status of the quiescent *Rhizoctonia* in soil and the latter method reflected the active status.

1. Immersion-tube technique (Mueller and Durrell, 1957).

The detailed procedure for isolating fungi by immersion tubes is described in Chapter 2. According to Martinson and Baker (1962) *Rhizoctonia solani* is easily isolated with such conventional laboratory media in the tubes as potato-dextrose agar or Modified Richard's agar (Chapter 16). Exudates from germinating radish seeds, combined with water agar or potato-dextrose agar, may significantly increase the frequency of isolation of this fungus.

2. Debris-particle method (Boosalis and Scharen, 1959).

A 100-g sample of soil is thoroughly suspended in about 2.5 liters of tap water in a metal vessel. The suspension is left to stand about 30 sec before the supernatant liquid is poured onto a sieve with openings of 250 μ. Soil that settles is resuspended in 1 liter of tap water and left to settle for about 30 sec and the supernatant liquid is passed through the sieve. This procedure is repeated 5-8 times, until the supernatant liquid is relatively free of plant debris particles. Seeds of weeds, large particles of soil and sand, and large pieces of straw and woody plant tissue are removed with forceps from the plant debris screening.

Most of the soil deposited on the screen is removed by holding the screen under running water and sharply tapping the frame of the screen against the sink. The remaining plant debris is spread on sheets of filter paper and left to dry for about 1 hr before placing on agar medium.

Water agar, adjusted with phosphoric acid to pH 4.8, is poured into petri dishes and inverted for at least 72 hr before use to eliminate moisture condensation on the surface of the medium. Four drops of streptomycin sulfate solution (20 mg/ml) are placed equidistantly on the medium in each dish. Approximately 1 hr after adding streptomycin, 1 plant debris particle is placed on each of the 4-areas of the medium treated with the antibiotic solution. The dishes are incubated at 24 C. Colonies of *R. solani* from particles are discernible in 4-96 hr, the majority developing in about 48 hr. Several species of fungi, other than *Rhizoctonia*, may grow out of the plant debris particles at first. These other fungi apparently do not prevent the slower-starting strains of *R. solani* from growing out of its natural substrate.

3. **Stem-colonization technique (Papavizas and Davey, 1959; 1962).**

Internodal stem pieces, 5-8 mm long, of mature, dry stems, such as buckwheat, cotton, or bean, are mixed with soil in suitable containers in the laboratory. Two g of stem segments are mixed with the equivalent of 300 g of oven-dry soil. The moisture content of the soil may be adjusted to 30-50% of the moisture-holding capacity. The soil-stem segment mixture is incubated at 23-24 C for 3-4 days.

After incubation, the segments are recovered by hand and washed for 20 min in running tap water. The washed segments then are tranferred to sterile paper towels, partially dried, and transferred to petri dishes containing 10 ml of water agar (2% agar). Fresh solutions of aureomycin hydrochloride, neomycin sulfate, and streptomycin sulfate are added to the water-agar medium at the rate of 50 ppm of each, after the medium has been melted and cooled to approximately 48 C. After the dishes are incubated at 23-25 C for 20-24 hr, they are placed under a microscope and examined for *Rhizoctonia*-type mycelia. The incubation period of the segments on the agar medium should never exceed 24 hr, since identification and isolation then would be hindered by saprophytic nematodes and secondary fungal colonizers spreading out from the segments. Hyphal tips of suspected *Rhizoctonia* mycelia are transferred to appropriate agar media.

K. *SCLEROTIUM*

1. *S. rolfsii* **(Leach and Davey, 1938).**

Sclerotia can be recovered by washing soil samples through a series of screens with openings of 2 mm, 841 μ, and 420 μ. All the sclerotia (except occasional aggregates) pass through the 2 mm screen. After the finer soil particles are washed through the screens, the residue from the 841- and 420-μ screens are flushed into a large white porcelain pan with sufficient water to make a depth of 2 cm. Under a strong light, sclerotia can be readily distinguished from weed seeds and other extraneous material and removed with forceps. To determine viability, the sclerotia are placed (without chemical treatment) on the surface of finely screened, unsterilized peat soil in petri dishes. The soil is then thoroughly moistened and incubated at 30 C for 5 days. The mycelium of *S. rolfsii* that develops can be readily identified on the background of black soil.

2. *S. cepivorum* (McCain, 1967).

Infested soil is dispersed in water by stirring for 30 sec in a blender at a slow speed and sieving through 595- and 250-μ sieves. The 595-μ fraction is discarded and the 250-μ fraction is treated with 0.525% sodium hypochlorite for 10 min. The hypochlorite aids in breaking up some of the soil aggregates which interfere with observation of the sclerotia. The hypochlorite also bleaches dark organic matter much more quickly than it does sclerotia, which are then readily observed. The sample is again dispersed in water in a blender for 10 sec, and passed through sieves of 500, 420, 350, 297, and 250 μ. The fraction from each sieve is examined under low magnification and the sclerotia are picked out with forceps. Viability of sclerotia is determined by placing them in 0.525% sodium hypochlorite for 2 min and plating on PDA.

L. *STREPTOMYCES SCABIES*

1. Selective medium (Menzies and Dade, 1959).

Soil dilutions are made and samples are plated on tyrosine-casein-nitrate agar medium (Chapter 16). Growth of bacteria is inhibited on this medium and a dark brown pigment closely encircles colonies of the pathogen. Almost all pathogenic isolates of *S. scabies* produce pigment, and most nonpathogenic actinomycetes do not; this reaction aids in the selection of the probable pathogen from dilution plates containing other morphologically similar actinomycetes.

2. Use of streptomycin-resistant strains (Weber et al., 1963).

Direct assays and population fluctuations of *S. scabies* in the soil can be made by using streptomycin-resistant strains of the organisms as diagnostic markers. Selection of streptomycin-resistant mutants of the organisms can be made with the gradient-plate technique (see Chapter 13). Resistant colonies are transferred successively to more concentrated gradient plates until a maximum or required level of resistance is achieved. Soil is artificially infested with spores of a resistant strain, and, after application of the treatment desired, surviving organisms may be recovered by plating soil dilutions on nutrient agar containing the maximum concentration or streptomycin tolerated by this strain.

M. *THIELAVIOPSIS BASICOLA*

1. Soil-dilution techniques.

The selective medium devised by Tsao (1964) for recovery of *T. basicola* from soil is a modification of Martin's rose bengal agar (see Chapter 16 for ingredients). Dilutions in water are made with sieved soil obtained adjacent to infected plants. The final dilution should be determined by trial, but Tsao found that 1:5,000 or 1:10,000 was satisfactory. One-ml samples of the final dilution are pipetted into petri dishes and liquid, but cooled, medium is added. The dishes are incubated in the dark at 25 C for 4-10 days. Colonies are examined directly under the microscope for endoconidia and chlamydospores characteristic of *T. basicola*.

Another selective medium was developed by Papavizas (1964) and designated as VDYA-PCNB agar. The medium contains 3 antibiotics plus oxgall and pentachloronitrobenzene (PCNB). See Chapter 16 for ingredients. Soil dilutions are made by suspending the equivalent of 5 g oven-dry soil in 500 ml sterile tap water and comminuting the suspension in a blender for 1 min at 2,000 rpm. Dilutions are made from the suspension while in motion on a magnetic stirrer. Final dilutions should be determined by trial but Papavizas found that 1:10,000 and 1:1,000 were satisfactory for rhizosphere and non-rhizosphere soil, respectively. Prepared dishes containing 1 ml of the final dilution in the selective medium are incubated at 25 C for 6-7 days. Colonies typical of *T. basicola* are counted and may be transferred to pure culture.

2. Carrot-disc technique (Yarwood, 1946).

Soil is spread over the surface of 5-mm-thick carrot discs in petri dishes and enough water added by atomizing to make the soil quite moist but with no free water present. After 2-4 days at room temperature, the discs are washed free of soil and reincubated in moist chambers. When soils containing *T. basicola* are used, grayish colonies appear in about 6 days after inoculation. At first, masses of endoconidia are formed, and later the colonies turn almost black as macroconidia are formed in abundance. Transfers of aerial mycelium can be made to tubes of potato-dextrose agar. To reduce the possibility of invalid results due to previous infection of the carrots with *T. basicola* or *Chalaropsis thielavioides*, the whole carrot should be surface-sterilized for 5 min in 0.5% sodium hypochlorite and washed with sterile water before being cut into discs (Lloyd and Lockwood, 1962).

N. *VERTICILLIUM ALBO-ATRUM* AND *V. DAHLIAE*

1. Soil dilution.

An alcohol-based selective medium was developed by Nadakavukaren and Horner (1959). Water agar (7.5 g agar/liter) is prepared and 90-ml portions are dispensed into 200-ml Erlenmeyer flasks. They are autoclaved and cooled to 42-44 C in a constant-temperature bath. Just before the diluted soil samples are mixed with the agar, a concentrated solution of streptomycin (to yield 100 ppm) and 0.5 ml of absolute ethyl alcohol are added to each flask. One ml of each diluted soil sample is added to each flask, and the contents are shaken and dispensed into petri dishes. They are incubated 10 days at 18-23 C in darkness. On this alcohol agar *Verticillium* colonies produce black microsclerotia, which can be counted easily against a white background. Most other fungi are colorless and grow sparsely. *Verticillium* colonies can be distinguished microscopically from other fungi that produce black mycelium.

An Anderson air sampler can be used to deposit small quantities of soil on alcohol agar plates (Harrison and Livingston, 1966). The authors suggest that use of the air sampler is more accurate than water dilutions of soil.

A dilute soil extract agar medium is useful for isolation of *Verticillium* (Menzies and Griebel, 1967). Ingredients in the medium as modified by Green and Papavizas (1968) are listed in Chapter 16. Growth of other fungi on this medium is very sparse, but *V. dahliae* and *V. albo-atrum* produce small heavily pigmented colonies that contain abundant microsclerotia. When the incubated plates are examined from the bottom by reflected light on a white background with a dissecting microscope at 7X magnification, *Verticillium* can be enumerated accurately and rapidly.

2. Soil washing (Evans, Snyder, and Wilhelm, 1966).

Microsclerotia of *Verticillium* are larger and heavier than most conidia in soil and may be separated from them with a soil-washing procedure. Small samples (50 mg) of finely sieved soil are shaken with 180 ml of sterile water containing 1% Calgon (sodium hexametaphosphate) in 200-ml flat medicine bottles. The suspension is left to stand for 5 min, and the supernatant is drawn off with a transfer pipette, leaving the residue in 20 ml of liquid. This decanting procedure is repeated 5 times with sterile water. One-ml portions of the final residual material are distributed over each of 20 petri dishes containing potato dextrose agar (5 g dextrose/liter and 200 mg of streptomycin/liter). The medium is prepared and poured 5 days previously to insure a dry surface. The dishes are rotated to distribute the suspension over the surface and incubated at room temperature for 7-14 days. Colonies of *Verticillium* can be identified with the aid of a dissecting microscope. Dilution plates prepared from each original 50-mg sample of soil are treated as one sample. The dilution factor of soil to water is 1:400.

A soil-washing tube later designed by Evans et al. (1967) facilitates the processing of samples in a shorter period of time.

CHAPTER 6
ISOLATION OF PLANT PATHOGENS FROM ROOTS

Some causal agents of diseases remained unknown for many years despite their common occurrence in plants or soil. As pointed out by Warcup (1959), this reflects a fault in procedure, particularly failure of workers to make visual and microscopic examinations of specimens for fruiting structures or mycelium that may not be obtained subsequently by the isolation procedure used. The most common procedure for isolating fungi from plant tissue is to surface disinfect (or disinfest) the plant part, cut small portions of tissue from the advancing margin of a lesion, and place these on a nutrient medium. For bacteria, maceration of the plant material, followed by dilution and streaking on a medium may be necessary. The kind of chemical disinfectant used, length of time tissue is treated, and kind of medium used will vary with the nature of the host tissue, the kind of pathogen suspected of being present, and personal preference of the operator based on his experience. Commonly used disinfectants are 0.1% mercuric chloride, 0.1% silver nitrate, 0.35% sodium or calcium hypochlorite, and 70% alcohol. Others used for various purposes are 1% carbolic acid, 5% formalin, 3% hydrogen peroxide, 0.1% mercuric cyanide, and 2% potassium permanganate. Most of these treatments should be followed by one or more sterile-water rinses. Plant parts with vascular infections and woody plants may sometimes be surface treated rather drastically without harm to the internal pathogen. Woody roots are frequently dipped in alcohol and flamed, the surface area is cut away with a flamed knife, and the tissue beneath is plated. Chemical disinfectants are not satisfactory for isolating certain fungi that are sensitive to such agents, or for treating small roots, or for tissues in advanced stages of decay. Serial washing with water may be adequate or desirable for surface cleaning of these materials.

Studies with wheat pathogens exemplify the influence of various disinfecting agents. Isolation of fungi from wheat roots was satisfactorily accomplished by Mead (1933) after treatment with fresh calcium hypochlorite for 20 min or

73

more, whereas mercuric chloride and silver nitrate were not satisfactory. Davies (1935) found that mercuric chloride, though satisfactory for isolation of *Helminthosporium sativum* and *Fusarium* spp., allowed other organisms to overgrow the relatively slow-growing *Ophiobolus graminis*. On the other hand, successful isolation of the latter was achieved after a 2-min treatment of plant material with 1% silver nitrate followed by rinsing in a sterile solution of sodium chloride.

Kind of culture medium and the incubation temperature are important factors in separation of organisms. Secondary organisms in tissue may grow more rapidly than pathogens on certain media. Snyder and Hansen (1947) discussed the advantages of using natural media and environments in the culture of fungi. An effective medium for isolation and sporulation of certain fungi was pea-straw agar prepared from water agar and chopped pea straw previously sterilized by fumigation with propylene oxide (Hansen and Snyder, 1947). Many other media (see Chapter 16), adjusted to a suitable pH or containing antibiotics for bacterial suppression, have been used for general and specific isolations. Incubation temperatures between 20 and 28 C are suitable for most pathogenic organisms, but some such as species of *Phytophthora* may require a temperature below this range, while *Sclerotium bataticola* and certain others grow better at slightly higher temperatures. The fact that some fungi can tolerate very low or high temperatures and continue to grow slowly while others are suppressed is an aid in isolation. Isolation can be facilitated if one is thoroughly familiar with the nutritional requirements of the pathogen and its degree of tolerance to extremes of pH, temperature, light, and moisture.

A. SPECIFIC PLANT PATHOGENS

Most fungi can be isolated from plant roots with conventional tissue-plating procedures and standard culture media. Therefore, we shall be concerned here primarily with fungi and bacteria which have been relatively difficult to isolate, though some of these might be obtained with standard procedures.

1. *Agrobacterium.*

A selective medium and a procedure for isolation of *A. tumefaciens* was devised by Patel (1926). Ten g of sodium taurocholate and 2 ml of a 1:1000 aqueous solution of crystal violet per liter are added to peptone-dextrose agar. Small pieces of gall tissue are finely chopped in 5 ml of sterile water in a sterile dish and left for 2-12 hr to free bacteria from the tissue. Then 1-2 ml of the infusion are placed in 50 ml of the melted agar, shaken well, and plates are poured. The plates are incubated at 27-30 C. Most bacteria other than the pathogen are inhibited either by the sodium taurocholate or the dye, and the crown-gall bacterium develops rapidly. Thus the necessity for drastic disinfection of tissue is avoided.

Ark and Thompson (1961) used a bait method to trap *A. rhizogenes* from infected tissue. Fleshy roots that can be used as baits are carrot, turnip, table beets, parsnip and Jerusalem artichoke. Such roots are washed in water containing a suitable detergent, then rinsed several times in distilled water and dried on paper towels. The roots are sliced longitudinally or cut into discs with a sterile knife and placed in deep petri dishes. Diseased parts of infected plants are ground in sterile distilled water and the suspension is applied with a cotton swab to the freshly cut surfaces. A profusion of roots will develop on the discs within 7-14 days. Sometimes both *A. rhizogenes* and *A. tumefaciens* appear on the same bait, but they can be separated with dilution or streaking techniques.

2. *Aphanomyces.*

A baiting technique can be used to trap zoospores of *A. euteiches* from infected pea roots (Sherwood, 1958). Root systems are selected from plants with wilted lower leaves. A 1-2-cm length of the upper primary root, with lateral roots attached but trimmed back, is washed in 8 changes of sterile water. The root is placed along the inside edge of a petri dish containing sterile water. Several corn kernels, previously autoclaved in water for 15 min, are arranged in the dish opposite the infected pea root. Zoospores are discharged from zoosporangia produced on the root, and secondary zoospores reach the corn kernels within 4-12 hr. Infected kernels are transferred to other dishes of sterile water; the water is changed every 8-12 hr to reduce bacterial growth. After 5-8 days the small mycelial mat growing from each kernel is removed with a portion of the kernel, rinsed in 3 changes of sterile water, blotted between sterile filter papers and plated on nutrient agar.

3. *Fusarium.*

Fusarium can be isolated from banana plant tissue with the method of Stover and Waite (1953, 1954). Pieces of washed and surface disinfected stalk or roots 2-5 cm long, or 1-5 g of small roots, are blended for 1-2 min in 100 ml of water. Portions (0.25 ml) of the mixture are transferred to dishes of Martins peptone-dextrose agar (pH 6.3) containing rose bengal and streptomycin (Chapter 16). Colonies of *Fusarium* grow well on this medium. The procedure is suitable for isolating vascular wilt fungi from plants other than banana and for studying rate and extent of pathogen growth in plant tissues.

4. *Phytophthora.*

Many of the baiting techniques described in Chapter 5 are useful for isolating species of *Phytophthora* from infected tissue. One of the older methods is that of using apples as baits. Infected tissue is forcibly inserted into the flesh of apples, or a cylinder of the flesh is removed with a cork borer and diseased tissue is placed in the hole. The hole is then closed with adhesive tape.

Phytophthora spp. usually can be isolated from advancing margins of decay. Species of *Pythium, Rhizopus,* and *Mucor* sometimes grow faster in apples than *Phytophthora* and may prevent its isolation.

It is important to select newly infected tissue for isolation of *Phytophthora*. Isolation is easily accomplished when roots of alfalfa infected with *P. cryptogea* are dipped in 70% alcohol, blotted dry, and tissue pieces from margins of affected areas are plated on water agar or cornmeal agar. Potato-dextrose agar is not satisfactory, nor is the fungus readily isolated from plants 1-3 years old (Erwin, 1954).

P. fragariae can be isolated from strawberry roots when a very short section of stele with a transition of color is embedded in water agar, or in oatmeal agar, as the medium is about to solidify (McKeen, 1958). Sporangia can be obtained in 24 hr when "tip" roots of infected plants are immersed in water.

A selective medium containing mixtures of certain antibiotics improves the recovery of *P. citrophthora, P. parasitica* and *P. cryptogea* from infected citrus and alfalfa roots (Eckert and Tsao, 1962). Root pieces 1 cm long are surface sterilized for 1-1.5 min in 0.5% sodium hypochlorite (10% Chlorox), rinsed in water and plated on cornmeal agar containing 100 ppm of pimaricin and 50 ppm each of penicillin and polymyxin. *Pythium* is usually the only other fungus that will grow on this medium.

5. *Pythium*.

A method described by Drechsler (1929) consists of placing roots suspected of harboring *Pythium* or other phycomycetous parasites in petri dishes containing sterile water. Where dense bacterial contamination is evident, the water is changed several times until it is no longer turbid. Within 12-24 hr, mycelium should appear and, after adequate development of the fungus, pieces of tissue are removed from the dish and blotted between filter papers. The pieces are transferred to a fresh medium, such as cornmeal agar. After growth of *Pythium,* hyphal tips can be transferred to fresh medium.

A medium and technique devised by Sleeth (1945) facilitates isolation of *Pythium* from diseased roots. The medium consists of (in g/liter of distilled water): dextrose, 10; $NH_4H_2PO_4$, 2; KNO_3, 1; $MgSO_4$, 1; agar, 25; pH 5-5.5. The medium in petri dishes is cut into 4 equal quarters with a sterile scalpel, and a small bit of tissue placed on each quarter. Three of the quarters are carefully lifted out with a flexible sterile scalpel and inverted in separate petri dishes. The fourth section is inverted in the original dish. Hyphae of *Pythium* grow through to the agar surface in 24 hr, usually in advance of other fungi and bacteria.

A method used by Johnson et al. (1969) has been successful for isolating *Pythium* spp. from diseased roots of sugarcane and corn, and from necrotic lesions on cotton hypocotyls. Necrotic portions (0.5-1.0 cm) of the roots or hypocotyls are cut from young plants and washed in gently flowing tap water for 24 hr. After they are washed in 3 successive changes of sterile water, each segment is placed on the surface of 2% agar containing 10 mg/liter of chlortetracycline (Aureomycin) in a petri dish. *Pythium,* if present, will usually be the first fungus to grow from the segment. After 2-3 days of incubation at 24 C, hyphal tips can be transferred to appropriate nutrient media.

6. *Streptomyces scabies.*

Contaminating fungi and bacteria are suppressed when scabby tissue is treated with phenol prior to isolation of the pathogen (Lawrence, 1956). Small portions containing lesions are removed from tubers and placed in a mortar with 20 ml of 1:140 phenol (v:v). The material is macerated with a pestle and the suspension is left to stand for 10 min, then one drop is spread over the surface of Czapek's agar medium at pH 6.5. With this method a large number of actinomycete colonies are obtained and these colonies are mostly free from contaminating fungi and bacteria.

A selective indicator medium developed by Menzies and Dade (1959) permits easy identification of *S. scabies.* Since it is difficult to distinguish the pathogen from saprophytic actinomycetes, verification of identification usually depends on conducting pathogenicity tests. A tyrosine-casein-nitrate medium (TCN) selects probable *S. scabies* isolates on the basis of their ability to produce a brown to black melanin-type pigment. Few isolates of the pathogen fail to produce this pigment, while saprophytic actinomycetes produce it very infrequently or not at all. Ingredients of the medium are found in Chapter 16.

7. *Thielaviopsis basicola.*

In a procedure described by Gilbert (1926), seed of tobacco are sown thickly on blotting paper moistened with sterile water in small glass moist chambers. Tiny bits of brown infected portions of diseased tobacco roots are scraped off and scattered among the tobacco seed. In a few days at the proper temperature the seed germinate and become infected with the mycelium of the pathogen. Infected seedlings are removed with forceps and plated on carrot agar. The fungus grows out readily and pure cultures are obtained by transferring.

Tsao and Van Gundy (1960) used a carrot technique adapted from that used by Yarwood (1946) to isolate *T. basicola* from soil. Surface-sterilized root pieces are inserted into narrow slits made on cut surfaces of carrots and incubated under moist conditions for 5-7 days. With this method the extent of field infection of citrus roots in southern California was studied. One should make appropriate tests to assure that carrots obtained from commercial markets are not already infected with *Thielaviopsis.*

8. *Verticillium albo-atrum.*

A sand-culture technique was devised by Wilhelm (1956) primarily for the purposes of disclosing limited infection in root systems and detecting symptomless carriers of *Verticillium albo-atrum* and other root-invading pathogens. Large quantities of roots or nearly entire root systems may be surveyed, whereas conventional culturing would require excessive time and labor. A quantity of previously washed and surface-disinfested roots are buried 5 cm deep in large petri dishes containing coarse, sterile, moist sand. The roots should be white, and without secondary tissues that may become dark colored. Immersion in 0.1% mercuric chloride for 1.5-2 min followed by 2 rinsings in sterile distilled water is adequate for disinfestation. Petri dish lids containing a layer of 2% water agar are used as covers to prevent contamination and loss of moisture from the sand. After 2-4 weeks incubation, roots are removed and examined under a dissecting microscope for resting or reproductive structures of fungi. These may be dissected out and mounted for critical examination, or transferred to appropriate culture media. If properly surface-sterilized before burial, roots will not deteriorate much during the incubation period. Certain fast-growing fungi may be trapped out of the sand during the first few days by placing sterilized seeds or bits of plant material near the roots.

B. WOOD ROTTING FUNGI

A general isolation procedure consists of washing pieces of infected woody roots, treating them with a disinfectant, rinsing them in sterile water, and plating small chips on nutrient media. Handling of some woody materials may be facilitated by use of special equipment, such as chisel forceps (Hubert, 1953), particularly where large numbers of isolations are to be made. A metal C-clamp with a jaw space of 8 cm or greater is fastened to a board about 15 cm wide, 13 cm long, and 2 cm thick, and fitted with a metal guard to prevent sliding on the table. Blocks of infected wood about 8 cm long are washed, dipped into boiling water (or disinfected as otherwise desired), then split with a sterilized blade to expose 2 inner uncontaminated surfaces. A piece selected for isolation is clamped in the metal jaws and fragments of wood are cut and removed with sterilized chisel forceps and transferred to the culture medium.

When attempting to isolate pathogenic Basidiomycetes from infected tissue, rapidly spreading mold contaminants, such as *Trichoderma viride,* often overgrow the pathogen. This difficulty was overcome by Russell (1956) who developed a medium selective for Basidiomycetes. The active selective ingredient of the medium is o-phenylphenol which inhibits growth of *Trichoderma, Penicillium, Cladosporium,* and other contaminants. Bacteria are inhibited by lowering the pH of the medium to about 3.5 with lactic acid. *Fomes annosus* and other wood rotting Basidiomycetes grow satisfactorily on this medium; the ingredients are listed in Chapter 16.

Rishbeth (1950) used a paper-wrap procedure for isolation of *Fomes annosus.* Wood samples are cleaned, wrapped in newspaper, thoroughly wetted with tap water, and incubated in glass containers. Presence of the fungus is readily detected by production of conidial heads of the imperfect stage

(Oedocephalum lineatum). Conidia usually appear in 4-5 days at 17-20 C but may require 8-10 days if tissues are resinous. If desired, isolates can be obtained from the conidia. Since the conidia are not readily confused with those of many other common wood-inhabiting fungi, this direct incubating method is reliable and time-saving. Fewer isolations on media are required to determine the extent of infection, and the common problem of fast-growing saprophytes on malt agar is avoided.

Hendrix and Kuhlman (1962), investigating the extent of wood penetration by *F. annosus*, compared the paper-wrapping technique with plating of chips on various agar media. They found that the paper-wrapping technique is suitable for determining presence or absence of *F. annosus*, but was not accurate for studies of infection, penetration and movement of the fungus in woody tissue. Chip isolating was particularly effective when used with PCNB agar, a selective medium (Kuhlman, 1966). See Chapter 16 for ingredients.

C. MYCORRHIZAL FUNGI

Generally, the mycorrhizae are thought of as comprising symbiotic relationships between fungi and plant roots, though Garrett (1956a) has pointed out that the difference between mycorrhizal fungi and highly specialized root-disease fungi may be one of degree rather than of kind. Since mycorrhizal fungi are not a separate taxonomic unit in the classification of fungi, methods of isolation and study are similar to those for other fungi. While isolation is sometimes difficult, the real problem in dealing with mycorrhizal fungi is the recognition or proof of a symbiotic relationship. Isolation is a necessary first step toward determining whether the relationship is one of symbiosis or of parasitism.

1. Root selection and washing (Robertson, 1954).

During investigations on the pattern of development of mycorrhizal roots of *Pinus sylvestris*, a suitable isolation method was sought for reliable diagnosis in field samplings. One method was based on selection of roots in the proper stage of development. In mycorrhizal work two kinds of roots are often referred to—long lateral roots and short roots. The short roots, which are produced on the long roots, typically develop into the familiar forked mycorrhizae. Prior to 1954, infection of the short roots was generally assumed to come only directly from fungi in the soil. Robertson observed that long roots in Scots pine also regularly become infected by a Hartig net without undergoing morphological change and that this infection may extend to the short roots. Long roots, the outer layers of which are still relatively smooth and not sloughing excessively, should be selected. These are thoroughly washed in a strong jet of tap water for 1 hr or more, then rinsed 3 times in sterile water and cut into pieces about 1-2-mm long. The pieces are plated on Martin's soil-extract glucose agar with yeast extract (0.05%), streptomycin (15 μg/ml), and rose bengal added. Plates are examined at intervals of 12 hr and hyphal tips of all fungi transferred to tubes of 2% malt agar. The combination of root selection and thorough washing provide root portions that usually are infected by only one fungus.

2. Section-embedding technique (Harrison, 1955).

The phycomycetous mycorrhizal fungi are usually seen as coenocytic hyphae penetrating cells of the outer cortex of roots. In the inner cortex they are mainly intercellular with arbuscules (haustoriumlike structures) developing within the cells. Terminal vesicles may form in some intercellular spaces, particularly in the outer cortex. Harrison devised a method for isolating *Pythium ultimum* from infected roots of *Allium ursinum*. Roots are thoroughly washed in a stream of tap water and transferred to a flask of sterile water where they are vigorously shaken for a few min. After 3 or more changes of sterile water longitudinal sections are cut aseptically. The sections are placed on sterile cover slips and covered with a few drops of 2% water agar containing 0.2 ml of Aureomycin per 15 ml of medium. The coverslip is then sealed onto a glass ring, as in the preparation of a hanging drop chamber. Daily observations are made and intercellular hyphae that grow from the cut edges of sections are transferred to water agar or malt agar. The washing process, quite similar to that used by Robertson (1954), is effective in removing most contaminating rhizosphere fungi. Aureomycin satisfactorily inhibits bacteria.

3. Pre-embedding sand-wash (Waid, 1957).

This method is essentially like that of the preceding except for a modified washing procedure. Roots are scrubbed with a test tube brush in running tap water and transferred to a 250-ml flask with 100 ml of sterile water and a little sterile sand to act as abrasive. The flask is shaken on a mechanical shaker for 10 min, then roots are transferred to a flask of sterile water without sand and shaken again for 10 min followed by 2 additional washings in sterile water. This treatment frees roots of external hyphae and spores, leaving only a few bacteria which are later controlled by Aureomycin in the section-embedding agar.

4. Ultrasonic bath treatment (Zak, 1962).

Successful direct isolation of fungal symbionts from pine mycorrhizae may be achieved by employing a detergent solution and an ultrasonic bath to clean roots. Following this treatment the roots are surface disinfected with sodium hypochlorite and thoroughly rinsed in sterile distilled water. Severed mycorrhizae are then plated on nutrient agar media.

D. VIRUSES

In virus studies the meaning of isolation may vary with the viewpoint of the researcher and the purpose of his investigation. To some, isolation may involve an extraction, chemical precipitation, ultracentrifugation and purification. Such procedures have been thoroughly treated in books on virology and in many research papers. For our purpose we shall regard isolation as meaning extraction of the infective principle free from host extaneous matter for the primary purpose of detection and determination of relative concentration prior to more detailed assay and study.

Since viruses in roots often affect aerial parts of plants also, in fact may have reached the roots via leaf infection, local lesion assay of root extracts on leaves is frequently applicable for detection in roots. Such procedures are rather standard and details can be found in virology texts. In many instances infected plant parts are simply triturated in a buffer solution using a sterilized mortar and pestle or a Waring Blendor and applied directly to indicator plants without further purification.

The following two specific techniques may relate to studies in soil microbiology.

1. Release of virus from roots (Yarwood, 1960).

In studies of factors involved in the infection cycles of TNV and TMV, plants may be grown in small pots with drainage holes and placed in larger waterproof glazed pots. The criterion for release of virus from the roots is detection of virus in the drainage water of plants with infected roots and absence of virus in water from noninfected plants. Assays of the drainage water and of extracts from roots are made on appropriate test plants. Some viruses apparently are released from roots of plants while others are not.

2. Recovery of TNV by *Olpidium brassicae* (Teakle and Yarwood, 1962).

The usual method for detecting tobacco necrosis virus is to macerate the roots of infected plants, dilute with K_2HPO_4 solution, and rub the suspension on cowpea leaves previously dusted with Carborundum. Since this technique has frequently failed to detect the virus in inoculated plants, another technique is suggested. Naturally infected roots may contain *Olpidium,* a known vector of TNV as well as TMV. Infected roots are washed and immersed in water into which zoospores of the fungus then become discharged. Mung bean seedlings are immersed in the zoospore suspension, then drained and incubated in moist chambers. Lesions usually develop on the roots in 2 days. Several days later the bean roots are assayed on cowpea leaves in the usual manner. This technique has shown TNV to be more common in crop and weed plants than previously shown by the direct root assay method. Success of the technique depends upon ready release of zoospores by the fungus. Under certain environmental conditions *Olpidium* may produce inactive resting spores instead.

CHAPTER 7

CONTROL OF SOIL ENVIRONMENT

Physical and chemical factors of soil environment which affect activities of soil microorganisms include moisture, temperature, atmosphere (oxygen and CO_2), nutrients, and pH. Some methods for controlling these factors are described in this chapter. Also included is a section on methods for soil sterilization.

A. SOIL MOISTURE

For microbial studies under laboratory conditions, soil is sifted through a screen with openings of 1 or 2 mm, and the moisture content is determined by drying a weighed portion for 24 hr at 110 C. The oven-dry soil is weighed and % moisture based on wt of dry soil is calculated by the following formula: $\frac{\text{wt of water}}{\text{wt of dry soil}} \times 100$. Soil can be brought up to any desired moisture level by atomizing and mixing weighed amounts with water. Wetted soil samples should be reweighed to determine the exact moisture content after spraying. To minimize moisture loss by evaporation during the spraying procedure, the required amount of water can be sprayed onto soil in a rotating drum. Another method involves adding to the soil the required amount of water in the form of crushed ice. The soil should be previously chilled so that the ice will not readily melt before it is mixed thoroughly with the soil particles.

To maintain a relatively constant moisture level, soil can be stored in polyethylene bags in the laboratory. Such bags are permeable to CO_2 and certain other gases, but only slightly permeable to water vapor. Kaufman and Williams (1962) found that the amount of water loss from soil in polyethylene bags at 27 C was less than 10% during a 6-week period. Microbial studies in the laboratory can be carried out in such bags, or in beakers covered tightly with polyethylene film.

In a review of terminology, measurement, and control of soil moisture, Couch et al. (1967) pointed out some of the problems of experimental control of soil moisture for greenhouse studies. Although it is a well-known fact that an intact soil mass cannot be wetted uniformly to stresses below Field Capacity (FC), some researchers continue to state in publications that pots of soil were watered periodically to maintain a certain moisture level. If a container is filled with dry soil having a field capacity of 30% and enough water is added to wet the whole mass to 15%, one-half of the soil will be wetted to field capacity and the other half will remain dry. In attempts to overcome this difficulty various workers have used double-walled pots, or they have added water through pipes buried to different depths in the soil. In each case, only a portion of the soil is wetted to FC; the remainder of the soil system is unchanged. Devices have also been proposed that supply water under tension or stress. Basically these devices consist of soil or sand columns of different heights above a reservoir of free water. Replenishment of moisture in the zones of active root extractions is accomplished by capillary rise. Variation in stresses in these areas is facilitated by the different heights above the water reservoir. It has been shown that movement of capillary water in a soil system is so slow that the moisture in the active root zone is extracted at a rate much faster than it can be resupplied by capillarity. Therefore, varying the height of a soil column works only for moisture levels above FC and cannot be considered valid for continuous stress studies in the readily available range.

Moisture control for short-term experiments in the greenhouse may consist of varying only the initial soil moisture content. Dry soil is brought up to desired moisture levels by atomizing the required amount of water onto soil particles in a rotating drum. No water is added to the intact soil mass during the course of the experiment. This method is useful for study of damping-off and seedling blight diseases.

1. **Varying frequency of irrigation.**

For long-term experiments, additional water must be added to the soil. Soil moisture variables are introduced by allowing the plants to extract to predetermined levels between Field Capacity (FC) and Permanent Wilting Percentage (PWP). The soil is then irrigated back to FC. The points between FC and PWP to which the soil is allowed to reach are determined by weighing periodically the entire soil-plant-container system. With this technique, arbitrary designations of moisture levels may be made. Thus, Johnson (1962) maintained moisture in 6-inch pots at ranges of 80-100, 50-100, and 25-100% of FC, reflecting high, medium, and low moisture levels, respectively.

2. **Split-root technique.**

Couch and Bloom (1960) utilized the principle that plant extraction of moisture is eventually stabilized at a point near PWP, and applied a modification of the split-root technique to permit study of living root systems near FC and at continuous PWP. Six-week-old plants (tomato) are transplanted to 2-compartment nonporous containers as follows: The root system is divided into 2 equal portions on the basis of orientation of the long axes of the stem. The basal end of the stem is then split longitudinally for 25 mm. One-half of the root system is placed in either side of the compartmented container (Fig. 7-1). Thus, the plant is supported by 2 root systems each dependent on separate irrigation. The soil of both compartments is held near FC by frequent irrigations until roots permeate the soil, then one root system is allowed to extract to PWP while the other is maintained near FC by frequent waterings.

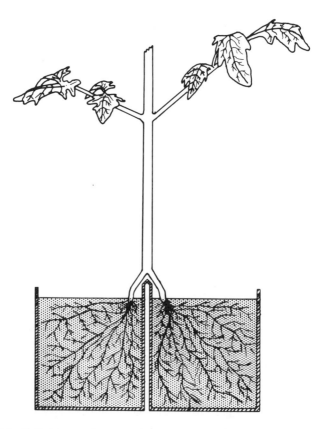

FIG. 7-1. Split-root technique. One root system is held at continuous PWP by withholding irrigations; the other root system is maintained near FC by frequent irrigations. (Couch and Bloom, 1960. Phytopathology 50:319-321.)

B. SOIL TEMPERATURE

Plant growth environmental chambers are often used for microbial and small plant studies in the laboratory. Diurnal programming of temperature, light, and humidity are available on expensive models. For most greenhouse studies, soil-temperature tanks are often used. They are basically of two types. In one type each tank is a unit in itself, having its own cooling unit and/or heating coils. Each is provided with a stirring apparatus or reciprocal water pump for agitation of the water. The other type involves use of separate cold water reservoir and hot water reservoir tanks. Cold or hot water is pumped from these to the different units, and water is returned to them from the units by gravity. The temperature of each plant growth unit is regulated by a thermostat; when either cold or hot water is needed by a particular unit, its thermostat activates a relay which opens an inlet valve on the tank and starts the pump from the proper reservoir supply.

For construction and details of popular types, see Campbell and Presley (1945), Ranney (1956), Cooper et al. (1960), Ferris et al. (1955), Dimock (1967), Harrison et al. (1965), and Steele (1967).

C. SOIL ATMOSPHERE

Soil pores that are not occupied by water are filled with gases which constitute the soil atmosphere. Respiring plant roots and microorganisms remove O_2 from the soil atmosphere and release CO_2 into it; thus, CO_2 concentration is higher than that above soil and the O_2 concentration is lower. There are many fundamental procedures and commercially available instruments for manipulating and measuring gases for various biological studies. Consult *Manometric Techniques* by Umbreit et al. (1964) for detailed procedures.

Soil gases have a direct effect on survival and activity of soil microorganisms. They also influence severity of root diseases. The literature on sampling and analysis of soil atmospheres and on the influence of CO_2 and O_2 on growth of microorganisms is voluminous. The descriptions below may serve to illustrate general principles of methods of study of soil atmosphere-microorganism relationships.

1. Sampling and analysis of soil atmospheres.

Gas samples are usually collected by means of various types of metal or glass tubes inserted in the soil. The technique of Yamaguchi et al. (1962), which is designed for sampling small volumes of gas, illustrates the general principle of gas collection with tubes. The samples (as small as 1 ml) are analyzed by gas chromatographic techniques. The sampling technique is a modification of the method used by Shapiro et al. (1956) to effect O_2 diffusion through soil. The gas collection tube (Fig. 7-2) is made by joining 6- to 8-mm tubing, 5 cm long, to

1.5-mm capillary tubing of desired length. The end of the 6- to 8-mm tubing is flared to make a base with diam not exceeding 2.5 cm. A short piece of tygon tubing containing a circular silicone gasket 1 cm in diam is attached to the upper end of the capillary tubing. The flared end of the apparatus provides a reservoir for gas equilibration at a given location in the soil. The total volume of the system is approximately 3 ml for a collection tube 25-cm long. A 2.5-cm core is removed from the soil, the tube is inserted, and soil is replaced and tamped to prevent gas leakage. Following an equilibration period, gas samples are drawn from the collection tube by means of a 1-ml syringe with a 22-gauge, 5-cm hypodermic needle. The silicone gasket can withstand repeated puncturing. After the gas sample is taken, the needle is withdrawn and inserted into another silicone gasket to plug the point. Samples can be held in this manner for at least

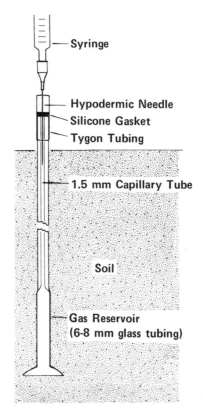

FIG. 7-2. Diagrammatic sketch of soil-atmosphere sampler. (Yamaguchi, 1962. Reprinted from Vol. 26, SSSA Proceedings pages 512-513, 1962, and reproduced with permission of the publisher.)

an hr without compositional change. A 0.05-ml quantity of the gas is expelled before injection of the sample into the gas chromatograph. In this case a modified Aerograph Model A-90C gas chromatograph was used and data recorded on a strip chart recorder with a 1.0-millivolt range. Accuracy of the sampling system should be tested with a gas mixture of known percentage concentration of CO_2 before being applied to an experiment. Such a gas mixture may be introduced at the bottom of a pot of quartz sand containing the tubes and let diffuse upward. Analysis of the gas collected should agree approximately with the composition of the aerating mixture.

Other apparatus and procedures for measuring both CO_2 and O_2 content of soil are described by Van Bavel (1965).

2. CO_2 effects on *Rhizoctonia solani* (Papavizas and Davey, 1962).

For studies of the inhibitory effect of CO_2 on *Rhizoctonia*, nonsterilized soils are permeated by atmospheres of various CO_2 concentrations. Commercial mixtures of CO_2 and atmospheric air containing 10, 20, and 30% CO_2 (v/v) are used, and atmospheric air alone serves as control. Cylinders with the desired proportions of CO_2 are connected by high-pressure-reducing valves to 1-liter Erlenmeyer flasks containing 750 ml of distilled water (Fig. 7-3). The gas mixtures are passed through the distilled water to humidify them. The main exit tube from the water flask is divided into several rubber side-arms of equal lengths, and each side-arm is connected to the inlet tube in a rubber stopper fitted into the low end of a glass tube (30 $\times$ 4.5 cm) containing sieved soil. Sterile glass wool 4 cm thick is placed between the stoppers and soil to facilitate gas diffusion. Replicate glass tubes of each treatment are mounted on a ring-stand. The upper end of each glass tube contains a No. 10 stopper provided with an outlet tube. Air is drawn through the atmospheric air control tubes by a vacuum pump. Flow rate of all mixtures is adjusted to renew the atmosphere in each tube every 2 min. Gas is collected from the tube outlets and analyzed (in this case with a Haldane apparatus). Effect of CO_2 on the competitive saprophytic activity of *Rhizoctonia* can be assessed by determining colonization of organic substrates incubated in the soils. The same apparatus may be used also in pathogenicity tests, where surface-disinfected seeds are planted in the top 1-2 cm of soil and emergence determined after exposures to gas mixtures.

3. CO_2 and O_2 effects of *Phytophthora* (Dukes and Apple, 1965).

The following procedure was employed to study the effect of gaseous mixtures on inoculum potential of *P. parasitica* var. *nicotianae* in two tobacco soils. Soil is screened to remove debris, and then fumigated with methyl bromide. The soil is then aerated, the moisture adjusted as desired, and uniformly infested with whole oat inoculum of the pathogen originating from a single-zoospore isolate. A quantity (14 liters) of the infested soil is placed in a

FIG. 7-3. Apparatus for studying the saprophytic and pathogenic activities of *Rhizoctonia* in unsterilized soil as affected by various concentrations of CO_2. (Papavizas and Davey, 1962. Phytopathology 52:759-765.)

sterile, large-mouth, 20-liter carboy; an equal volume to serve as a check is treated the same way but not enclosed in the carboy. The carboy is closed with a rubber stopper previously wrapped with aluminum foil and coated with petrolatum to prevent loss of CO_2. One short and one long glass tube extend through the stopper, the long tube extending to within 2.5 mm of the bottom to permit injection or removal of a gas mixture without opening the carboy. These tubes outside the carboy are fitted with polyethylene tubes which are closed with two screw-type clamps each. Initially, an analysis is made of the O_2 and CO_2 content of the soil air, then a gaseous mixture (i.e., 5% O_2, 15% CO_2 and 80% N_2) is injected into the soil in the carboy at near-atmospheric pressure. Flow of this gas mixture is allowed until the escaping gas is the same composition as the test mixture. Analyses of O_2 and CO_2 are made at 24-hr intervals for 5 weeks, and the gases are replaced when their concentration varies more than 2% from the test mixture. The carboy with the soil is stored side down in the dark at a suitable temperature (26-30 C) and rotated each day to prevent accumulation of gases in certain areas.

At termination of the 5-week period the soil is placed into 30 10-cm pots properly replicated and planted with tobacco plants (resistant and susceptible to black shank). Extent and rapidity of disease development is then calculated by a disease index of 0 (no disease after 20 days) to 20 (all plants diseased after 2 days).

D. SOIL FERTILITY AND pH

Control of soil fertility and pH is often attempted in plant disease studies. It may be desirable to determine the effect of concentration of major and minor nutrient elements and their interaction with pH on inoculum potential and disease development. Such a study in a field test becomes largely a problem of agronomic management. The reader is referred to a comprehensive text on soil fertility such as that of Tisdale and Nelson (1966). It should be pointed out that soil in the experimental area should be described prior to starting such an experiment. The description should include soil type and series, C:N ratio, pH, and initial concentrations of nutrient elements to be manipulated. Standard and accepted methods of soil analysis should be used. These can be found in the American Society of Agronomy publication, "Methods of Soil Analysis" I and II, edited by C. A. Black, 1965.

In the greenhouse, soil pH may be adjusted by the following method (Johnson, 1962). A soil, such as Hartsells loam, which has a naturally low pH of about 4.5, is selected. Hydrated lime is mixed with samples of the soil. The amount of lime to add to yield a certain pH will depend on the buffers present and exchange capacity of the soil and should be determined previously by trial. A maximum pH of about 8.0 can be obtained with large amounts of lime. Levels above this can be obtained by adding powdered NaOH. Before use, the lime-amended soil should be incubated in the greenhouse for about 2 months for the pH to stabilize. If soil with a low pH level cannot be obtained, or if there is reason to use a certain soil with a pH of 6-7, the pH may be raised by adding NaOH and lowered with H_2SO_4. A dilute solution of H_2SO_4 can be atomized onto the soil in a rotating drum to ensure thorough mixing.

When studying the effect of nutrient elements on inoculum potential or disease development, one must keep in mind that changes in these elements can take place quickly in soil. Thus, ammonia in soil might be converted to nitrate and some lost by volatilization. Nitrate is easily leached from soil or reduced and lost as nitrogen gas. Various chemical tranformations can occur with other nutrient elements. To minimize such changes, nutrition studies are often performed on plants grown in sand culture, such as those by Bloom and Couch (1960) and Ghabrial and Pirone (1966). Consult the original articles for detailed procedures.

E. SOIL STERILIZATION

When soil is used only as a substrate to study growth, survival, and pathogenesis of a single pathogen or of known mixed isolates subjected to various treatments, complete sterilization is desired prior to introducing the organisms. In some instances, only partial sterilization is needed to free the soil of unwanted pathogens while leaving a reduced population of the natural microflora for studies of the competitive nature of pathogens. Each sterilization method (heat, chemical, or irradiation) has certain disadvantages, foremost of which is the alteration of soil from its natural state.

1. Sterilization by heat.

Soil may be treated with a dry source of heat from flame or electricity, but this is far less desirable than steam heat because of the high temperature required and poor distribution of heat through the soil. Steam provides a large quantity of heat at low intensity, and flows through the soil to all areas. The phycomycetous fungi are most sensitive to heat. Free-flowing pressureless steam is used in most greenhouses because of economy of equipment and convenience of applying heat to stationary soil beds. Many pathogenic fungi, along with nematodes, insects, and most weed seeds, are killed by a temperature of 83 C for 30 min and pressureless steam.

Steam under pressure is most often used in laboratory work and certain greenhouse work for sterilizing soil in movable containers. Such treatment is provided with autoclaves heated by gas or electricity, or supplied with steam directly from a central source. Autoclaving is usually at 15 lb pressure with a temperature of 121 C (250 F). Many workers tend to autoclave soil longer than is necessary. A treatment of 30-40 min, then again after about 24 hr for 30 min usually is sufficient. Certain factors will influence the efficiency of sterilization. Heat plus moisture is more effective in killing microorganisms, but excessive moisture may prolong the time necessary to reach the proper sterilizing temperature and may increase the base exchangeable ammonia. Length of required heat treatment increases with weight and depth of the sample and with increased amount of organic matter. Soil should be free of lumps before sterilizing. A reduction of nitrate in soil and an increase in uptake of certain minerals by plants is sometimes experienced with autoclaved soil. See Baker and Roistacher (1957) for further discussions of sterilizing equipment and the principles of heat treatment.

Partial sterilization of soil at a relatively low temperature may be achieved by using aerated steam (Baker and Olson, 1960). The standard temperature for steaming soil is 100 C for 30 min, but steam temperature may be lowered to any desired level by dilution with air. The rate and distance of movement of aerated steam through soil approximates that of pure steam. Treatment with aerated steam may be surface or subsurface application, or in a steam vault with a moving soil mass. Aeration is effected by a steam jet exhauster or a squirrel-cage blower. Low-temperature steam has certain advantages: application cost is reduced; a smaller boiler is required, or a larger area may be treated in a given time; post-steaming soil toxicity is decreased; and a portion of the non-pathogenic natural soil microflora remains, providing competition for pathogens that may be introduced. While the primary purpose of this procedure is to rid soil of pathogenic organisms so that plants can be safely grown, it may be applicable to other experimental studies as well.

2. Chemical sterilization.

The application and efficacy of volatile chemicals, such as chloropicrin, methyl bromide and ethylene dibromide, for reducing populations of microorganisms in soil is discussed thoroughly by Munnecke (1957). These materials are used to treat soil without heat in soil flats, soil bins, and small field plots. Certain common fungicides such as vapam and terraclor (PCNB), and nematocides, also are used for soil treatment in some experiments.

Allison (1951) described a procedure for vapor-phase sterilization with ethylene oxide. Soil is air-dried and sieved through 2- or 3-mm openings, then moistened to about 2/3 FC and placed in convenient containers inside a vacuum desiccator or wide-mouth jar. A tall chamber can be made by inverting one large desiccator over another. Vapor from a tank of ethylene oxide is drawn through the chamber by means of a pump or water aspirator. Less than one atmosphere of gas is usually sufficient for a period of from 6-24 hr for most soils. The gas was found to cause no increase in soluble salts, but water soluble organic matter was increased by adsorption of the vapors on organic and inorganic colloids.

Propylene oxide also is an effective gaseous sterilant (Hansen and Snyder, 1947). It is slightly less explosive than ethylene oxide.

3. Sterilization by irradiation (McLaren et al., 1962).

Irradiation of soil may be an improved procedure for differential sterilization to eliminate fungi, bacteria, and viruses without drastically altering soil structure. Radiation-sterilized soil still manifests certain enzyme activity (phosphotase, urease) in presence of suitable substrates, is not toxic to tomato plants, and does not provide extra nutrient to plants. The principal disadvantage of the technique at present is the unavailability of necessary energy sources and trained operators in many institutions. In part of the work of McLaren et al., a horizontal electron accelerator operating at 5 Mev. was used to treat soil samples on a turntable rotated at 20 rpm. The soil to be irradiated was sealed in polyethylene bags of 15 × 38-cm size. The bags were held taut and the soil was pressed against the turntable rim so that maximum thickness was 1.8-2 cm ensuring penetration of beam energy through the entire soil depth. The amount of dose delivered was monitored by measuring that fraction of electrons which was collected by a sheet of aluminum foil covering the bags and rim. Consult the original article for details of dose determinations and standard curves. Bacterial numbers in soils approached zero at 2 Mrep. doses, but 4 Mrep. doses were necessary for complete sterility of larger soil volumes. Survival curves for microorganisms in a given soil sample were very similar for soil treated with high energy electrons, gamma rays, or hard X-rays.

CHAPTER 8
EXTRACTION OF SOIL SOLUTIONS

The chemical and physical nature of the soil solution has a direct influence on growth and survival of soil microorganisms. This solution contains mineral nutrients and soluble organic materials required for metabolism and growth of both microorganisms and higher plants; also present are inhibitory constituents. Separating the liquid from the solid portion of the soil is often necessary for analyses of these soluble materials. Once separated, the liquid may be tested by standard biological procedures for its influence on growth and activity of soil microorganisms. Individual chemicals present in the soil liquid can be isolated by extraction and chromatographic techniques.

One should be aware of the limitations of the procedures for separating the solution from soil. An "average" soil solution is difficult to obtain. The liquid phase of the soil is heterogeneous with respect to the concentration of dissolved substances. Because of the presence of finely divided solid particles, dissolved substances are subjected to positive or negative adsorptive forces as their concentration decreases or increases. As the soil becomes dryer (either naturally, or as the result of solution extracting procedures) the concentration of dissolved substances may increase or decrease depending on the substance involved, the pH of the soil, and the soil structure. Thus, an extracted soil solution is not a true soil solution since the liquid closely bound to soil particles is usually not extracted. Moreover, the adsorptive forces acting on the dissolved substances in soil are not active in an extracted soil solution and this solution has little or no buffer capacity. It should be recognized that quantitative measurements of exchangeable cations cannot be obtained through use of the extraction procedures described below. The methods will provide, however, a measure of soluble salts and organic substances from samples of a soil provided that the same extraction procedure is used for each sample.

A. WATER EXTRACTS OF SOIL

To obtain a "dilute" soil solution, a volume of water is added to soil, the mixture is stirred, and the liquid material is filtered. Parker (1921) proposed that water be added to soil at the ratio of 5:1 by volume. The mixture is stirred for 3 minutes and left to settle for 12 min. The suspension then is filtered.

For use in culture media, microbiologists usually prepare "heated" extracts of soil. James (1958) found more bacteria isolated on dilution plates prepared with heated soil extract than on plates containing unheated extract. The procedure of James (1958) and of Allen (1957) are outlined in Chapter 16.

For obtaining soil solutions the simple method of making water extracts of soil has advantages, such as ease of handling and short time necessary for the extraction procedure; expensive apparatus is not required. But when solutions obtained in this manner are subjected to chemical and biological analyses, several limitations are apparent. There is some doubt that the method provides a true quantitative measurement of salts and other soluble materials. Addition of large quantities of water alters the equilibrium of the soil, and has a solvent effect that may cause precipitation of some of the materials in the soil solution due to an alteration in the nature of the solvent.

To obtain samples of a soil extract at any time during an extended period of extraction, a soil-perfusion apparatus may be used. An artificial biological system can be maintained during several days of perfusion, and soluble materials present in the perfusate can be analyzed chemically or biologically at periodic intervals. An illustration of the soil-perfusion apparatus as modified by Audus (1946) is seen in Fig. 8-1. Soil is placed in the glass tube P between pieces of glass wool, and the perfusion solution is placed in the separatory funnel F. A constant small suction is applied at A by a suction pump. This suction is transmitted back through the lengths of thermometer tubing R_1 and R_2 and the soil column in P to the perfusion solution in tube T. This causes air to be drawn in at the base of the side tube S, thereby detaching a column of solution, which is drawn up tube T and discharged on the top of the soil column. Upon release of the tension in T by this discharge, solution again rises above the base of S until it reaches the level of the solution in F. More air is now drawn in at the base of S and the whole process is repeated. The liquid discharged on top of the soil column is drawn with the air stream through the soil and drains back into F. The bypass tube with flow resistance R_2 is not necessary for the working of the apparatus, but is useful in regulating the degree of aeration of the soil. The air flow is divided between the soil column and this tube in the inverse ratio of the resistances of these two arms to air flow. Resistance of R_1 should be high in relation to the other resistances in the circuit to insure steady rates of flow of air and solution. T should be about 4-mm internal diameter. To sample the perfusate, the suction is released and the perfusate rises in side-tube S where samples can be pipetted off.

A soil-perfusion apparatus (Fig. 8-2) designed by Kaufman (1966) is equally adaptable to either positive or negative pressure systems. When the lower end of the delivery tube is in contact with the solution surface in the reservoir, the air

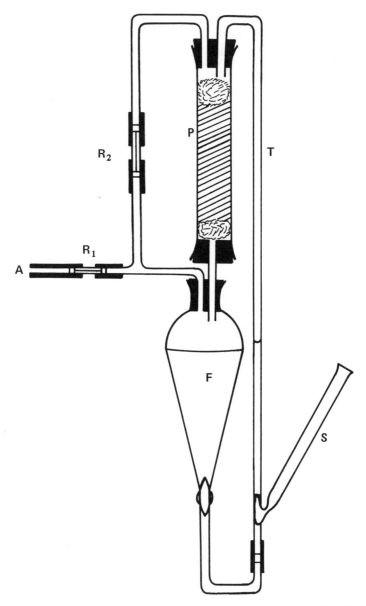

FIG. 8-1. Soil-perfusion apparatus. (Audus, 1946. Nature 158:419.)

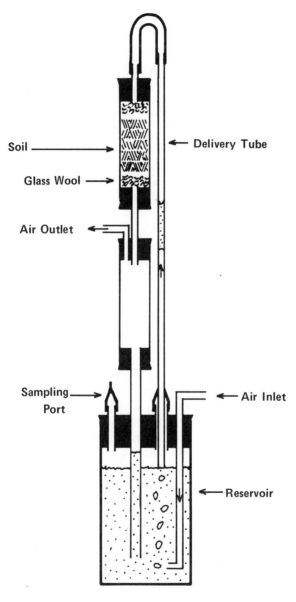

FIG. 8-2. Diagram of soil-perfusion unit. (Kaufman, 1966. Weeds 44:90-91.)

that enters via the air inlet lifts a detached column of the solution up through the delivery tube where it is discharged onto the top of the soil column. Solution perfused through the soil falls back into the reservoir, thus permitting the air, which also must pass through the soil column, to escape through the air outlet. The amount and rate of solution to be recycled, and the degree of aeration, can be controlled by the depth to which the delivery tube penetrates the solution in the reservoir and by controlling, externally, the rate and amount of air introduced into the system. Presence of a second tube beneath the soil column permits the buildup of pressure within the system whenever the soil becomes blocked. Under normal operating conditions (porous soil through which solution moves freely) very little back pressure exists in the system. Samples can be removed by inserting a pipette through the sampling port, which consists of a short piece of rubber tubing held shut with a small clamp. Opening of the sampling port releases the pressure exerted by the incoming air, thus stopping the recycling of the solution. When closed again the recycling process resumes automatically.

B. VACUUM FILTRATION

A widely used method of obtaining soil solution for soluble salt analyses is to subject water-saturated soil to vacuum filtration. Electrical conductivity techniques are then used to determine total amounts of salts in the solution obtained. Concentrations of different salts and ions in the solution may be determined by various chemical tests.

Essentially, the method consists of making a paste by adding distilled water to the soil sample with adequate stirring. The soil paste is placed on tightly seated filter paper in a Buchner funnel and the "saturation extract" filtrate is removed by suction (Jackson, M.L., 1958). Several modifications of the vacuum-filtration assembly have been suggested. Richards (1949) described the use of plastic filter funnels (commercially available) to be substituted for Buchner funnels. Sheard (1961) described a plastic filtering jar assembly that permits direct filtration into a volumetric flask. A semi-micro filtration assembly utilizing commercially available components was devised by Boswall and Mackay (1963). A suction-plate apparatus (Fig. 8-3) was designed by Leo (1963). The metal or plastic perforated plate is placed on a funnel suspended in a flask with side arm. The soil extract is collected in a 35-ml plastic or glass vial within the flask. According to Leo, the apparatus is easier to clean than the conventional Buchner funnel and time required for extraction is reduced.

A technique was described by Reeve and Doering (1965) for obtaining samples of soil solution in the field. It consists of placing in the soil ceramic cups which are connected with neoprene tubing to a vacuum pump. Solution is collected in traps along the vacuum line. The technique is especially useful for collecting solution samples from soil during irrigation.

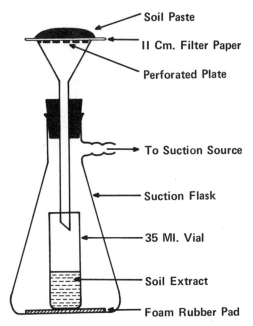

FIG. 8-3. Suction-plate apparatus. (Leo, 1963. Soil Science 95:142-143. © 1913, The Williams & Wilkins Co., Baltimore.)

C. CENTRIFUGATION

Solutions can be extracted from soil by centrifugation of soil pastes or of natural soil with a relatively high moisture content. The amount of extract obtained depends on the centrifugal force used, the time of centrifugation, and the moisture content of the soil.

1. Centrifugation of soil pastes (Chesnin and Johnson, 1950).

Soil pastes are prepared and centrifuged at 2,400 rpm for 10 min. To remove soil particles that might fall from the sides of the tube during decantation, the supernatant is filtered through the perforations in Gooch crucibles fitted with discs of filter paper and placed under suction. Previously, a drop of extract that had separated from the soil paste during centrifuging is withdrawn with a pipette and placed on top of the filter paper disc so that the paper is wet and firmly pressed in place. The remainder of the extract is decanted from the tubes and poured through the filter. Saturation extracts collected in this manner are clear and free of colloids. However, if the soil is highly colloidal or dispersed, higher centrifuge speeds may be necessary to obtain clear separation. An average recovery of about 30% of the water used to prepare soil pastes can be obtained.

Extraction of soil solutions

2. Centrifugation apparatus (Davies and Davies, 1963).

Small samples of solution can be extracted from natural soil with this apparatus. Rigid Perspex (plexiglas) weighing bottles are used. They are 5.0 cm high, with a rim 1.0 cm from the open end where the external diameter is 2.6 cm. Two bottles (Fig. 8-4) are required. In the bottom of one is bored a central hole of about 0.3 cm diam; the rim of the second bottle is removed, and the rough edges are made smooth with fine sandpaper. Both weighing bottles are placed in a polyethylene centrifuge tube (10.0 cm high, 2.7 cm internal diam) with a rounded bottom which is packed with small pieces of filter paper. The unbored weighing bottle (a) is inserted first into the centrifuge tube and rests on the paper packing; the bored bottle (b) is then pushed in. It is held there partly by its rim resting on the top of the containing centrifuge tube and partly by support from bottle (a). Bottle (b) is closed by a cap having two small holes bored in it. The whole assembly is then fitted into a standard metal centrifuge shield. Before use, the inside bottom of bottle (b) is padded with a small piece of glass wool. Soil is placed on the glass wool and tamped down with a glass rod. Bottle (a) is weighed and the apparatus is fitted together and balanced against another centrifuge tube containing water. Samples prepared in this manner are centrifuged for at least 60 min at 3,000 rpm giving a relative centrifugal force at

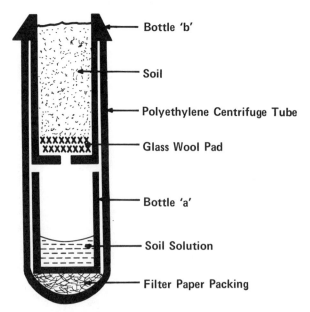

FIG. 8-4. Centrifugation assembly. (Davies and Davies, 1963. Nature 198:216-217.)

the center of gravity of 1,200 G. After centrifugation, bottle (a) is removed and weighed. The amount of solution removed can be compared to the total moisture content determined by oven drying another sample.

D. DISPLACEMENT WITH ALCOHOL (PARKER, 1921)

According to Parker, a soil solution obtained by displacement with alcohol approached being a "true" soil solution with respect to removal of soluble salts. Successive portions of the displaced solution contain the same amount of total salts. Also, the concentration of salts in the displaced solution is inversely proportional to the moisture content of the soil. The method is simple and can be used on all soil classes at a wide range of soil moisture content.

Moist soil is packed in a cylinder with an outlet at the bottom. The higher the soil column and the more compact the soil, the greater will be the percentage of soil solution displaced. High moisture content also tends to produce a high per cent displacement. Ethyl alcohol is poured on top of the soil column. When the saturated zone reaches the bottom of the soil column, the clear soil solution is free of the displacing liquid and drops from the soil as gravitational water. The time for displacement depends on the moisture content of the soil, degree of packing, soil structure and type, and height of soil column. A column 30-35 cm in height may require 12 hr for displacement to be completed. From 30-40% of the solution in natural soil can be displaced.

During the final stages of displacement the presence of alcohol in the displaced solution is usually noticeable by the turbidity of the solution and by the different index of refraction. The iodoform reaction may be used to determine presence of alcohol in portions of the displaced solution. A small crystal of iodine or a few ml of an aqueous solution of iodine-potassium iodide are added to the liquid to be tested. Then sufficient potassium hydroxide is added to give the liquid a distinct yellow to brownish color. The solution is warmed gently and, if alcohol is present, a whitish to lemon yellow precipitate of iodoform will appear (Scott, 1925).

E. PRESSURE-MEMBRANE APPARATUS (RICHARDS, 1941, 1947)

The pressure-membrane apparatus provides a convenient reliable means of removing soil moisture under controlled conditions from soil samples throughout the whole plant growth range, and without disturbance to the soil structure. The method may be used on prepared samples or undisturbed soil cores.

Soil is placed on a cellulose or "sausage casing" membrane inside a filter chamber (Fig. 8-5). Compressed nitrogen gas is applied in the chamber and water in the soil is forced through the microscopic pores of the membrane. This liquid is collected from an outflow tube into a beaker placed under the chamber. The

100 *Extraction of soil solutions*

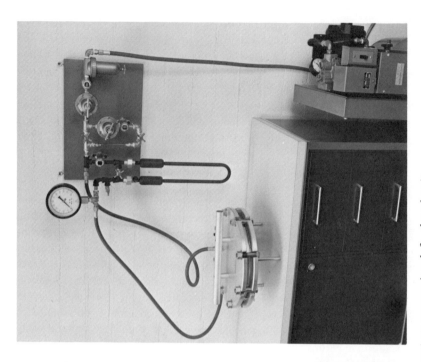
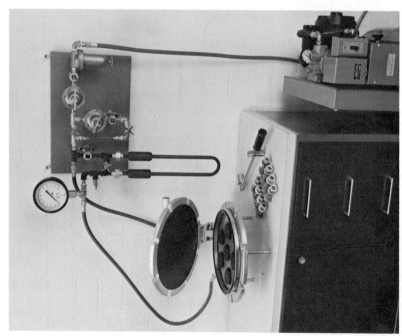

FIG. 8-5. Pressure-membrane apparatus: open unit at left, closed unit at right. (Courtesy Soil Moisture Equipment Co.)

moisture content of the soil in contact with the membrane will be reduced by the amount that would be necessary under normal atmospheric conditions to make the pressure deficiency of the soil water equal to the excess gas pressure in the extraction chamber. At equilibrium (when the flow of moisture ceases) there is an exact relationship between the air pressure in the extractor and the soil suction (and hence the moisture content) in the sample. If the pressure is maintained at 15 atmospheres (225 psi), the moisture held at this pressure is the approximate permanent wilting percentage (pwp).

CHAPTER 9
RESPIRATION AND ENZYME ACTIVITY

Physiological activities of soil microorganisms such as respiration and enzyme production and activity can be measured and related to certain growth phenomena in soil. When soil organic matter levels are increased, there is usually a corresponding increase in oxygen consumption, CO_2 evolution, total microbial counts, enzyme activity, and organic matter decomposition rates. Respiratory rates and enzyme activity, therefore, can be used as measures of saprophytic activity and colonization of crop residues by plant pathogens (Hine and Trujillo, 1966; Rodriguez-Kabana and Curl, 1968). Many bacterial and fungal cells exist in soil for long periods in a relatively quiescent state and respiration rates are low. For this reason, there is often no correlation between respiration rates and microbial population densities in soils not amended with organic matter.

A. RESPIRATION

A variety of respirometers and modifications of some of the earlier techniques provide useful means of measuring respiration. For the principles of manometric techniques, consult the manual of Umbreit et al. (1964). Many of the procedures and principles involved in measuring microbial respiration are outlined by Stotzky (1965).

1. Oxygen consumption.

Difficulties, not usually experienced in conventional experiments, were encountered by Chase and Gray (1957) who investigated the feasibility of using the Warburg respirometer to study microbial respiration daily in soil. Operation of the water bath at 5 C or more above room temperature for several days causes water vapor in the manometer to interfere with normal operations. This condensation can be avoided by operating the bath at 25 C. A 20% solution of KOH for CO_2 absorption in center wells of Warburg vessels is unsatisfactory because it extracts moisture from the soil. This is partly corrected by using a 3% solution of KOH, which then necessitates changing the solution at intervals of

2-6 days, depending upon rate of CO_2 evolution. The tendency of the solution to creep over the vaseline-coated rim of the well after several days can be overcome by using a paddle-shaped piece of filter paper in the well. The "blade" of the paper is of such a width that it touches opposite walls of the well and has a narrow handle extending up through the mouth but not touching the greased rim of the well.

For operation of the Warburg respirometer, samples (1-2 g) of soil are placed in the main chambers of conventional single side-arm vessels. The soil is distributed evenly on the vessel bottom and water is added to provide 60% water holding capacity. The rims of center wells are greased, and 0.2 ml of 3% KOH solution is placed into each well and the "paddle-shaped" filter paper is carefully inserted. The flasks are attached to the manometers and these are randomized for proper distribution of replicates. A thermobarometer consisting of a manometer with flask containing 2-3 ml of water is included, usually in duplicate. Temperature of the water bath is usually kept at 25 C and flasks are incubated without agitation. An equilibration period of 1 hr is allowed the first day and a period of ½ hr on each successive day before the stopcocks are closed. Manometers are read at 2-hr intervals for 6-10 hr of each day, though only the first and last of each set of daily readings may need to be used. From these, rate of O_2 uptake for each day is calculated and expressed in microliters per gram of soil per hour ($\mu l/g/hr$). Immediately following the last reading on any day, stopcocks on the manometers are opened and the water circulator turned off; the thermostatically controlled heater in the water bath is left on to maintain a uniform temperature overnight. The circulator is started the next morning, and after 30 min the manometers are closed and measurements started for another day.

Large Warburg flasks (125-ml) with center well, vented side-arm, and standard taper ground-glass neck may be used (Gilmour et al., 1958). This permits use of larger soil samples (50 g) than is possible with the standard 25-ml Warburg flasks. Two milliliters of 20% KOH are added to the center well and folded filter paper is placed in each well to absorb CO_2 evolved. Individual flasks are attached to calibrated manometers and placed in a constant temperature water bath (28 C). For each soil treatment, appropriate nontreated controls are included to reflect normal gas exchange. Oxygen consumption is expressed as microliters of gas consumed.

2. Carbon dioxide evolution.

a. **Continuous flow system.** Carbon dioxide evolution has been used more extensively than O_2 uptake as a measure of soil microbial activity. Apparatus and procedures have been modified so frequently to suit particular experimental objectives that few workers can agree that one method is better than another; however, the basic principles involved are similar.

Stotzky et al. (1958) described a large incubation unit for measuring CO_2 evolution and an additional unit in which subsamples of soil can be removed during incubation if desired. Procedures are clearly outlined and the apparatus diagrammed in detail in a more recent publication (Stotzky, 1965). Also described is a special apparatus for measuring the respiratory quotient of soil (RQ = CO_2 evolved/O_2 consumed).

Rodriguez-Kabana et al. (1967) found a basic CO_2 collection apparatus most suitable for studying the activity of pure cultures of *Sclerotium rolfsii* in flasks of sterilized soil treated with herbicides. A simple unit is diagrammed in Fig. 9-1. An extensive system of these can be connected in series and continuously aerated by means of small vacuum pumps. Construction is simple and utilizes components that are standard equipment in most laboratories. The scrubber system provides CO_2-free moist air that sweeps the duplicate soil flasks G and H. Air enters a 1-liter aspirator bottle (A), which contains (top to bottom) glass wool (a), Drierite ($CaSO_4$) desiccant with indicator dye (b), desiccant without indicator dye (c), and another layer of glass wool (a). The desiccant under average conditions should last about 1 week, then a color change will indicate that it should be replaced. Dry air from this bottle is essential before entering tube B. This is a glass tube, 4.7 cm (internal diam) × 70 cm, containing (top to bottom) glass wool (a), NaOH pellets (technical) (d), desiccant with

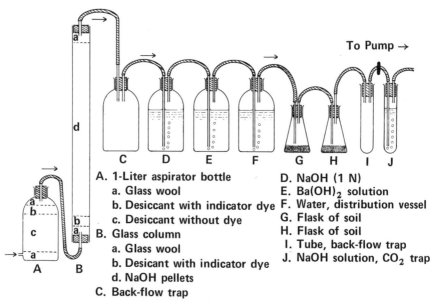

A. 1-Liter aspirator bottle
   a. Glass wool
   b. Desiccant with indicator dye
   c. Desiccant without dye
B. Glass column
   a. Glass wool
   b. Desicant with indicator dye
   d. NaOH pellets
C. Back-flow trap
D. NaOH (1 N)
E. Ba(OH)$_2$ solution
F. Water, distribution vessel
G. Flask of soil
H. Flask of soil
I. Tube, back-flow trap
J. NaOH solution, CO_2 trap

FIG. 9-1. Carbon dioxide-collection apparatus. (Courtesy R. Rodriguez-Kabana)

indicator dye (b), and another layer of glass wool (a). If moist air reaches the column, the NaOH pellets may coalesce and stop air flow through the system. This column should continue to remove CO_2 efficiently from the air for several months.

A 1-liter bottle (C) serves as a backflow trap to guard against changes in pressure, such as caused by electrical failure, that may flood the NaOH pellet column. Bottle D is ¾ full of 1 N NaOH which removes remaining traces of CO_2 and begins the air-moistening process. Bottle E contains a 1% (w/v) solution of $Ba(OH)_2$ to serve as a CO_2 indicator, and bottle F containing water completes the moistening process, after which the CO_2-free air sweeps duplicate soil flasks G and H. Erlenmeyer flasks of 250-ml capacity, with no more than 100 g of soil, are satisfactory. After sweeping over the soil, the air passes through back-flow trap I (glass tube, 32 × 200 mm), and CO_2 from the two flasks is trapped in NaOH solution in tube J (no more than ¾ full). The concentration of this solution will depend upon the amount of respiration expected; normality values between 0.1 and 0.5 usually meet requirements.

The tubing that joins I to J should be surgical tubing with a greased screw clamp attached. Air flow is regulated with the clamp at this point by adjusting the number of bubbles flowing from a 1-mm capillary tube. Capillary tubes of this size are essential, because larger diameters do not permit accurate regulation of air flow through the system. Rate of flow is usually adjusted to 15-20 bubbles/min. The air from tube J passes either through a desiccation step or directly into a vacuum pump. Small diaphragm pumps with maximum pressure of 18 psi and maximum vacuum of 18 Hf are inexpensive and can accommodate up to 40 individual CO_2 traps connected to a simple scrubber.

Either natural soil, or soil autoclaved and inoculated with a specific organism, may be used. One hundred-g samples of screened soil, with moisture adjusted as desired, are placed in a series of 250-ml flasks fitted with 2-hole rubber stoppers. A short piece of glass tubing filled with loosely packed cotton is inserted in each stopper hole to serve as filters when the flasks are attached to the vacuum assembly. If sterilization is desired, the flasks are autoclaved at 121 C for 1 hr, and again for 45 min 24 hr later. These are then inoculated aseptically with the desired organism and subjected to any treatments required in the experiment. Duplicate flasks (per treatment) are connected at positions G and H of the system, and the screw-clamp between I and J is opened fully. Two additional flasks with autoclaved noninoculated soil are connected also to serve as blanks. The pump is started and air is pulled through the system for 15 min; then the flow is adjusted to a desired rate with the screw-clamp.

Samples for CO_2 analysis are taken from tube J at desired intervals (usually 24-28 hr) during the incubation period. The screw-clamp should be closed, and the pump is turned off and left 5 min for the system to equilibrate. Then the NaOH tube J is opened and two 25-ml samples are transferred to 150-ml

Erlenmeyer flasks and closed with rubber stoppers. To each flask are added 50 ml of 1 N $BaCl_2$ solution and 0.2 ml of phenolphthalein solution. Each flask is titrated with standard 0.1 N HCl to the white endpoint, and results are averaged. The volume of NaOH solution remaining in tube J is recorded and discarded, and the tube supplied with fresh solution.

Calculations can be made as follows:

$$\text{Milliequivalents } CO_2\text{-C}/100 \text{ g soil} = \frac{(V^b - V^s)NV^t}{50}$$

where

V^b = volume (ml) of HCl to titrate alkali in CO_2 sample from control blank
V^s = volume (ml) of HCl to titrate alkali in CO_2 sample from soil treatment
N = normality of HCl
V^t = total volume in NaOH trap.

The calculations will give meq CO_2-C/100 g soil during the particular time interval covered by the sampling, or the sum of these interval values will give total CO_2-C evolved for the incubation period.

This CO_2 collection system, with properly placed screw clamps, can take as many as 60 soil-flask units at one time with considerable uniformity in flow between units. It is advisable to connect extra units with empty flasks at G and H throughout the system to safeguard against backflow. The main criticism of such gas trains is the problem of maintenance. The system must be checked frequently (twice each day) for proper flow performance of pumps and condition of desiccants.

b. **No-flow system for CO_2**. Simplified determinations of soil respiratory activity, such as those described by Elkan and Moore (1962) and Bartha and Pramer (1965), have been used to avoid gas trains and expensive Warburg-type respirometers.

The apparatus (Fig. 9-2) of Bartha and Pramer was made by Bellco Glass Co., Inc., Vineland, New Jersey. It consists principally of a 250-ml flask (H) fused to a 50-ml, round bottom tube (C). The flask is closed with a rubber stopper in which is mounted an Ascarite filter (F) provided with a stopper and stopcock (G). The side tube (C) is closed with a rubber stopper in which is inserted a 15-gauge needle (B) 15 cm long. The needle is capped with a rubber policeman (A) and its tip covered with a short length of polyethylene tubing (E) that touches the bottom of the side tube.

Fifty g of soil are placed in the flask (H), adjusted to 70% of moisture holding capacity, and the flask is closed. The filter stopper is removed and stopcock G opened. Side unit C is charged with alkali by injecting 10 ml of 0.1 N KOH through the opened canula (B) with a calibrated syringe. Stopcock G is closed, the syringe removed, and the rubber policeman and filter stopper are returned to initial positions. CO_2 produced by organisms in the soil is absorbed by the alkali and determined volumetrically.

To recover the alkali for analysis, the procedure for charging the unit is performed in reverse. Side tube (C) is rinsed with CO_2-free water and recharged with alkali by means of the syringe. The wash water and KOH are combined in a 50-ml flask with 1.0 ml of 2 N $BaCl_2$, and titrated with 0.05 N HCl to a phenolphthalein end point. Calculations are similar to those given for the continuous flow method.

To avoid contamination with CO_2, stock alkali and wash water may be stored in large bottles fitted permanently with Ascarite filters and syringe needles. This apparatus and method of analysis facilitate the monitoring of CO_2 production at frequent intervals for prolonged periods with the same soil sample.

Though the method is relatively simple it does require purchase or manufacture of a special apparatus. Since CO_2-free air is not made available continuously, oxygen may become limiting in rapidly respiring soils.

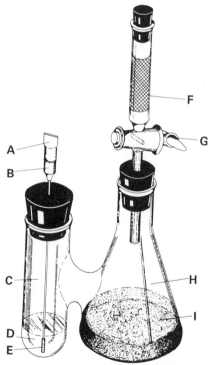

FIG. 9-2. Flask for monitoring the production of CO_2 in soil. (Bartha and Pramer, 1965. Soil Science 100:68-70. © 1965, The Williams & Wilkins Co., Baltimore.)

B. ENZYME ACTIVITY

Agricultural soils contain enzymes in amounts that can be determined with conventional methods of enzyme chemistry through action on appropriate substrates. Enzymes in soils, while sometimes associated with roots, are believed to be predominantly of microbial origin. Total dehydrogenase or catalase may provide information about the microbial population as a whole, while cellulase, saccharase, and xylanase may indicate activity of specialized microbial groups. It is advisable in natural soils, however, to consider enzyme activity *per se.* Thus, we would think in terms of what effect treatment A has on saccharase activity of soil C, rather than interpret enzyme activity as directly related to microbial population. For further general information the reader is referred to the reviews of Hofmann (1963), Porter (1965), and Skujins (1967).

Assays can be performed in either fresh or air-dried soils, though loss of some enzyme activity by air-drying samples at ordinary room temperature may occur. This can be minimized by freeze-drying the soil, but for routine processing of large numbers of samples it may be assumed that loss from air-drying will be constant in all samples and the activity remaining will be comparable. Storage of the dried soil in a cool (4 C) dry atmosphere is recommended. The samples may be placed in plastic bags and stored in sealed bottles containing a drying agent such as Drierite.

The amount of soil to be used will depend on the amount of activity present in the soil, and is determined by preliminary analysis with representative samples and varying amounts of soil (1-20 g). A linear relationship should be obtained between increasing amounts of soil and enzyme activity. If such a relationship is not seen, the amount of enzyme activity probably exceeds the capacity of the assay method; therefore, the amount of soil used must be decreased.

The following are selected examples of analyses of dehydrogenase and invertase (saccharase) and are based primarily on information provided by R. Rodriguez-Kabana.

1. Dehydrogenase activity.

Dehydrogenase activity in soil reflects metabolic rate and is not always correlated with microbial numbers in unamended soils, but with the addition of nutrients, activity generally increases with microbial numbers. Dehydrogenase activity without addition of glucose can be interpreted as the result of endogenous respiration in the soil.

The principle of the method is that formazan is formed by actively metabolizing cells when they come in contact with aqueous solutions of tetrazolium salts. Formazan is extracted and the intensity of the colored extract is used to assess the amount of metabolic activity (Lenhard, 1956, 1957). The procedure follows:

Soil (1-20 g, dry-wt basis) is placed in each of four 50-ml Erlenmeyer flasks. One of the flasks receives 5 ml of water, another 5 ml of Solution A, the third 5 ml of Solution B, and the fourth 5 ml of Solution C. Flasks are stoppered with 2-hole rubber stoppers with inserted glass tubing.

The solutions are as follows:

Solution A — 0.5% (w/v) glucose.
Solution B — 0.5% (w/v) solution of 2,3,5-triphenyltetrazolium chloride (TPC).
Solution C — 0.5% (w/v) solution of TPC and 0.5% (w/v) glucose.

The flasks are swept with N_2 for 10 min and the connecting tubes are closed with screw clamps leaving each flask with a N atmosphere. Flasks are placed in a water bath at 30 C and incubated for 1-6 hr, depending on the amount of dehydrogenase activity present in the soil. Occasionally the incubation period must be extended for soils with low microbial activity. The flasks are removed from the water bath and 10 ml of methanol are added to each, then they are swirled and filtered by gravity. This is repeated with successive 5-ml aliquots of methanol until all formazan has been removed (final volume, 55 ml). Occasionally some formazan is chromatographed on the filter paper, and care should be taken to wash it down the funnel with methanol.

The filtrates are made up to equal volumes with methanol and optical density is read in a suitable spectrophotometer using a wavelength value of 485 mμ. A standard curve is prepared: 5 mg of triphenylformazan (TPF) is dissolved in methanol to a volume of 500 ml. This stock solution is diluted to provide TPF amounts in the 0-1 mg/100 ml range. Dilutions are best performed using 100-ml volumetric flasks. These formazan solutions are read at 485 mμ and values used to prepare a standard curve relating O.D. to mg of formazan. Formazan values are converted to μl H (1 mg of TPF required 150.35 μl H for its formation). The O.D. value obtained for the water flask is subtracted from the O.D. value for the TPC flask, and the O.D. for the glucose flask is subtracted from that obtained for the glucose X TPC flask. The results are converted to μ moles H using the standard curve. Activity is expressed in terms of μ moles H/g of soil/hr.

The foregoing method is based on that described by Skujins and McLaren (1968). The original procedure requires the use of Thunberg tubes, for which we have substituted 50-ml Erlenmeyer flasks, and N sweeping for 10 min; the reaction requires anaerobic conditions. Two treatments are applied, one providing glucose as the energy source and the other without glucose, the latter depending on the energy sources already present in the soil. These two values should be considered in assessing dehydrogenase activity in soils. Prolonged incubation periods (12 hr or longer) should be avoided, since this may result in proliferation of microbial numbers even under anaerobic conditions. If activity is too low the amount of soil used should be increased. We have described the required treatments for a single replication. If conditions are well standardized a total of 4 replications should be sufficient for most soils.

2. Saccharase activity.

Saccharase activity in soils results in the hydrolysis of sucrose to its common components, glucose + fructose. Assay methods for these products are quite numerous and can be based on reductometric methods, polarimetry, and enzymatic determinations of glucose released. Described below is a reductometric procedure. The following solutions are needed:

Solution A: Aqueous solution containing 69.28 g of $CuSO_4 \cdot 5H_2O$/liter.

Solution B: Dissolve 103.2 g NaOH in about 600 ml of distilled water, then add slowly while stirring 346 g of sodium potassium tartrate; cool to room temperature and make up to 1 liter.

Solution C: Sulfuric acid and water (1:3 v/v).

Solution D: KI, 20% (w/v).

Solution E: 0.1 N sodium thiosulfate. Dissolve 250 g of crystalline sodium thiosulfate in 1 liter of water. Let stand for 72 hr and filter. Dilute to 1/10 in water and standardize the approximately 0.1 N thiosulfate solution using the dichromate procedure of Treadwell (1937) as follows: Make up a 0.1 N solution of $K_2Cr_2O_7$ by dissolving 4.903 g of the dry salt in water and dilute to 1 liter. Pipette 25 ml of the dichromate solution into a 500-ml Erlenmeyer flask containing 50 ml of water, 10 ml of concentrated HCl (sp gr 1.191) and 3 g KI. Allow the reaction to proceed in the dark for 5 min, dilute to 400 ml, and titrate with 0.1 N thiosulfate solution. Starch indicator (2 ml) should be added after 20 ml of thiosulfate have been added.

$$\text{Normality of thiosulfate} = \frac{2.5}{\text{ml of thiosulfate used}}$$

Starch indicator: Knead 0.5 g of soluble starch in a little cold water. Add 25 ml boiling water and simmer for 2 min. Cool to room temperature and add 1.0 g KI.

Solution F: Prepare a pH 5.00 acetate buffer by adding aqueous M-acetic acid (60.05 g/liter) to a molar solution of sodium acetate (136.08 g/liter) until a pH 5.00 is attained.

Solution G: Sucrose in water, 20% (w/v).

Two 10-g quantities (or less in case of high activity) of sieved soil (1-mm fraction) are transferred separately to 100-ml volumetric flasks. Control flasks with autoclaved soil should be run simultaneously to check for possibility of acid hydrolysis of sucrose. To prevent microbial growth, 2.5 ml of toluene are added to each flask. Each flask is shaken well without spattering soil against the flask wall, and let settle for 15 min.

To each flask are added 10 ml of Solution F, then 10 ml of water to one flask and 10 ml of Solution G to the other. These are shaken gently, closed with rubber stopper, and incubated at 37 C for the desired time (24 hr for most natural soils). The flasks are filled to the mark with warm water (37 C) and 2.5 ml additional water is added to account for the toluene volume. They are then shaken and incubated for another hour to permit soil particles to settle.

A 10-ml aliquot of supernatant is transferred from each flask to separate sugar tubes containing 10 ml of Fehling solution and 5 ml of water. These are boiled for 10 min in a water bath and cooled to room temperature. To each tube are added 4 ml of Solution C, 5 ml of Solution D, and 0.5 ml of starch indicator.

Titration is performed with standardized thiosulfate solution, and results expressed in terms of ml of 0.1 N thiosulfate solution required. The difference between values obtained for the flask with water and the flask with sucrose solution represents the saccharase activity of the soil. This is expressed in terms of ml of 0.1 N sodium thiosulfate/10 g of soil/24 hr. If desired, saccharase activity can be represented in terms of glucose or fructose released. To do this, a standard curve is prepared by adding to the tubes increasing amounts of either glucose or fructose in the range of 0-30 mg.

A buffer of pH 5.00 is suggested but it is recommended that a preliminary test be made with molar buffers at various pH values (acetate and phosphate buffers provide an adequate range) to determine the optimal pH for measuring saccharase activity of the soils in question.

The reductometric assay is well adapted for routine assays of most soils, but is not the most sensitive of the many Fehling tests in the literature. When poor soils are used it is recommended that a more sensitive test be used to analyze reducing sugars in the supernatant. Such tests are the volumetric method of Somogyi (1945) or the colorimetric procedure of Nelson (1944). The amount of soil should also be increased and the final volume reduced from 100 to 50 ml. Incubation time may also be lengthened for particularly poor soils.

3. **Technique for mixed fungal culture.**

Variables common to all microorganisms, such as CO_2 evolution, O_2 consumption, and total dehydrogenase activity, are useful indicators of total biological activity but do not allow for resolution of individual microbial responses in a mixed culture. Interactions between microbial species in soil can be studied, provided the enzyme systems of the contending organisms are different. Then with proper methodology a measure of specific enzyme activity of individuals in a dual-culture system under a particular set of conditions may provide a quantitative description of growth for each organism. A primary prerequisite, however, is a thorough knowledge of the enzyme system of each organism to be used in the mixture.

For this type of study certain preliminary tests must be made to assure that the enzyme activity of each organism does provide an accurate description of growth under the environmental conditions to be used in the experiment. For example, does the enzyme activity relate directly to mycelial dry weight and CO_2 production? No general procedure can be described because each mixture of organisms has its own peculiarities. We will use as an example the mixed culture experiments conducted with *Sclerotium rolfsii* (plant pathogen) and *Trichoderma viride* (antagonistic saprophyte) by Rodriguez-Kabana (1969). Only studies with saccharase will be discussed here. The following filter-sterilized nutrient solution is required (composition per liter):

Dextrose	30.0 g
K_2HPO_4 (dibasic)	1.0 g
$MgSO_4 \cdot 7H_2O$	0.5 g
KCl	0.5 g
KNO_3	2.0 g
Stock solution of $FeSO_4 \cdot 7H_2O$ (1 g/liter)	10.0 ml
Stock solution of $MnSO_4 \cdot 7H_2O$ (1 g/liter)	6.0 ml
Stock solution of $ZnSO_4 \cdot 7H_2O$ (1 g/liter)	10.0 ml
Stock solution of thiamine (1 g/liter)	1.0 ml

Adjust medium to pH 5.10 with 50% lactic acid

a. **Optimal pH for enzyme activity.** The first step is to determine the buffer pH necessary for maximal saccharase activity. Flasks of autoclaved soil, moistened with the nutrient solution above, are inoculated with the organism (*S. rolfsii* in this case). After 6 days at 28 C, soil from the flasks is air dried and saccharase activity determined using 0.1 M buffers at pH values ranging from 2.5-9.5 (see Table 1). Enzyme assays are performed at each pH; controls included should be soil with no substrate added and soil with no enzyme activity (autoclaved soil). From the results a curve like that in Fig. 9-3, can be constructed to show the pH for maximal saccharase activity. When the worker is dealing with other enzymes it is well to be aware that phosphates inhibit activity of phosphatase and also may interfere with the reagent used for assay of maltase activity.

b. **Standard curve for enzyme activity.** A standard curve for saccharase can be prepared as follows. Mycelium is harvested from 10-day-old cultures of *S. rolfsii* grown in Czapek's liquid medium containing 75 μg of thiamine hydrochloride per ml. Five g of the fresh mycelium are transferred to a cold blender with 25 ml of cold demineralized water. This mixture is blended for 45 sec, after which 225 ml of cold water are added and blended for 5 sec. Increasing volumes

TABLE 1
COMPOSITION OF SODIUM SALT BUFFERS USED IN ESTABLISHING RELATIONSHIPS BETWEEN PH AND ENZYMATIC ACTIVITIES. (RODRIGUEZ-KABANA, 1969)

pH	Ionic composition
2.5	Citrate
3.0	Citrate
3.5	Citrate-Acetate
4.0	Acetate
4.5	Acetate
5.0	Acetate or Citrate
5.5	Citrate or Acetate-Phosphate
6.0	Citrate
6.5	Maleate-Tris (hydroxymethylaminomethane) or Phosphate
7.0	Phosphate or Tris-Maleate
7.5	Tris-Maleate
8.0	Tris
8.5	Borate-Tris
9.5	Borate

of this suspension are then added to 20 g of autoclaved soil in 100-ml volumetric flasks and the soil is treated with toluene. Differences in volume are corrected by adding cold demineralized water so that the final volume in all flasks is the same. Four 10-ml aliquots of the chopped mycelial suspension are oven dried for 48 hr at 70 C in tared, perforated aluminum cups, and the mycelial dry weight per ml of the suspension is determined. Saccharase activity in the mycelium-treated soil is determined and a curve constructed relating this activity to mycelial dry weight (Fig. 9-4); this serves as a guide for further experiments. It is important to establish the enzyme X mycelium relationship in autoclaved soil that is identical to the soil to be used for culturing the fungus.

It is advisable to perform other preliminary tests to determine whether there is a linear relationship between soil enzyme activity and growth of the organisms being studied. Such a relationship for *S. rolfsii* is evident by its linear enzymatic response to increased concentration of nutrient added to soil and by the relation of saccharase activity to CO_2 evolved (Rodriguez-Kabana and Curl, 1968).

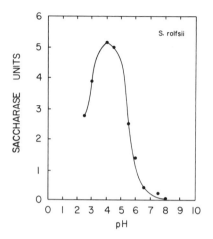

FIG. 9-3
Effect of pH on enzyme activity of soil colonized by *Sclerotium rolfsii*. (Rodriguez-Kabana, 1969. Phytopathology 59:910-936.)

FIG. 9-4
Relation of enzyme activity to amount of mycelial matter of *Sclerotium rolfsii* added to autoclaved soil (Rodriguez-Kabana, 1969. Phytopathology 59:910-921.)

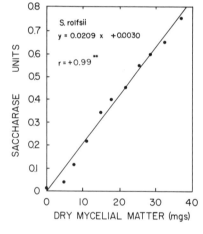

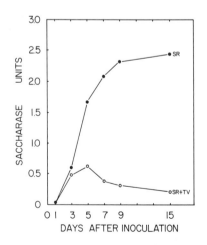

FIG. 9-5
Growth of *Sclerotium rolfsii* (SR) alone and in mixed culture (SR + TV) with *Trichoderma viride* as indicated by saccharase activity. (Rodriguez-Kabana, 1969. Phytopathology 59:910-926.)

Having thoroughly established these facts for individuals, one may then grow two organisms such as *S. rolfsii* (SR) and *T. viride* (TV) together in sterilized soil and obtain a measure of their interaction by assaying for activity of a particular enzyme of one that is not possessed by the other under a given set of experimental conditions.

    c. **Mixed culture.** A nutrient solution is added to flasks with 100 g (dry wt basis) of autoclaved soil so that each flask receives 360 mg glucose, 3.36 mg N as KNO_3, and 2.13 mg P as K_2HPO_4. The flasks are divided into 4 groups: SR, SR + TV, TV, and uninoculated (UN). The SR and SR + TV flasks are inoculated with 2 ml of chopped mycelial suspension of *S. rolfsii* prepared from young PDA cultures. The inoculum is applied by pipette in a straight line across the soil surface. Flasks in the TV and UN groups are left uninoculated at this time. Final moisture of the soil should be 60% MHC.

All flasks are incubated at 28 C for 24-48 hr to permit the pathogen to become established. Then the SR + TV and TV flasks are inoculated with 2 ml of a mycelial suspension of *T. viride*. In the SR + TV flasks this inoculum should be added at right angles to the previous pathogen inoculum. Five flasks from each treatment are now taken from the incubator and the soil is air dried in preparation for enzyme analysis. Other samples are taken at intervals throughout the 15-day period. The soil is analyzed for saccharase activity according to the procedure already described.

Flasks of soil with pure culture of *S. rolfsii*, supplied with a nutrient solution based on glucose, NO_3-N and PO_4-P, exhibit considerable saccharase activity, whereas similar soil from *T. viride* cultures does not. Fig. 9-5 illustrates the suppressive effect of *T. viride* (in the SR + TV culture) upon *S. rolfsii* as compared to uninhibited enzyme activity of the pathogen growing alone (SR culture).

In addition to enzyme activity, other chemical analyses may provide a better understanding of the mechanism involved in the biological control effect. Examples are residual glucose, NO_3-N, and P in the soil (indicating effect on nutrient utilization) pH, titratable acidity, conductivity of soil solution, etc. (see Rodriguez-Kabana et al., 1967; Rodriguez-Kabana and Curl, 1968). For procedures and references to analyses of soil enzymes other than saccharase, see Rodriguez-Kabana (1969).

CHAPTER 10

GROWTH AND SURVIVAL

To study growth and survival of an organism in soil is to become involved with the many complex interrelationships of biotic and physical soil factors that influence spore germination, substrate colonization, mycelial extension or cell multiplication, competitive activities, sporulation and sclerotial formation, fungistasis and lysis, and various other phenomena. Knowledge of the capacity of a soil to inhibit germination of spores or growth of hyphae of a particular fungus is needed for an understanding of its saprophytic and parasitic activity. A study of the influence of environmental conditions, organic amendments, and fungicides or bactericides on survival of pathogens is often a prerequisite for successful control measures.

The following methods were selected to represent the many original and modified procedures described for experimentation with various phenomena related to growth or survival.

A. ASSESSING GROWTH IN SOIL

1. Glass tubes.

Growth of *Sclerotium bataticola* through soil can be studied in glass tubes filled with soil (Norton, 1953). Tubes about 60 cm long (13 mm ID) are sealed at one end and filled to a depth of 40 cm with soil that has been previously air-dried, sieved through a screen with openings of 2 mm, and mixed with the desired amount of water.

The tubes are plugged with cotton and autoclaved for 2 hr if sterilization is required. An agar disc containing spores or mycelium of the test fungus is placed at one end of the soil column. When a food base is not desired, washed mycelium from a liquid culture is applied. Growth of mycelium through the soil column is observed with a dissecting microscope.

A soil-recolonization tube (Fig. 10-1) facilitates sampling of soil during incubation and appears to be adaptable to studies of recolonization, growth and survival of fungi, bacteria, and actinomycetes in variously treated soils (Evans,

1955). The apparatus consists of pyrex tubing 56 cm long and 2.5 cm ID; seven side-arms 2.5 cm long and 1.3 cm ID are fused into the main tube at 5.0-cm intervals. The tube is filled with sterilized or otherwise treated soil adjusted to a desired moisture content. The soil is evenly packed by lightly tapping after each addition of 25 g moist soil. The whole apparatus may be sterilized after the soil is placed in the tube. Nonsterilized soil, amended soil, or fungal inoculum may be placed in the end of the main tube. Ends of the main tube are plugged with cotton and those of the side-arms covered with metal caps. Water lost by evaporation can be restored periodically by weighing the tube and adding sterile distilled water through the side-arms. At suitable intervals, samples of soil are taken from the side-arms with a sterile microspatula, and soil dilutions or soil plates are prepared for determination of the microflora. A set of control tubes should be included.

For purposes of studying the distribution of fumigants, fungicides, or other materials in soil, the ends of the main tube and of the side-arms are sealed with rubber stoppers after the apparatus has been filled with soil of proper moisture content. The material is introduced through one side-arm and the apparatus is resealed for the desired period. The rubber stoppers are removed and replaced either with cotton plugs or metal caps. At intervals the distribution of the material is determined by isolating soil microorganisms from microsamples taken from the side-arms, and by comparing them with samples taken from untreated control soil tubes.

2. Glass slides.

A glass-slide technique developed by Blair (1942, 1945) is an adaptation of the Rossi-Cholodny buried-slide technique to study growth of fungi in soil. Chemically cleaned glass slides are placed vertically or radially (on edge) in soil in glass tumblers or 10-cm clay pots. The soil should be previously screened and the moisture adjusted as desired. An agar disc of the fungus is placed against each slide at a marked point. The soil and slides are then covered with sand (moisture adjusted as for the soil) to a depth of 1.3 cm. As the fungus grows some hyphae come in contact with, and grow along, the slide surface. This is

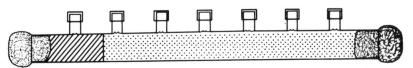

FIG. 10-1. Soil-recolonization tube. Hatched area, nonsterilized soil; stippled area, sterilized or treated soil. (Evans, 1955. Transactions of the British Mycological Society 38: 335-346.)

taken as an indication of the extent of spread of the fungus through the soil as influenced by particular soil conditions. The slides are removed at intervals, separated from the soil, fixed, and stained in a steaming solution of 5% erythrosin in 5% carbolic acid. Fungal strands on the slide are observed with a microscope and extent of growth is measured.

3. Plate procedure (Vujicic and Park, 1964).

The influence of food base on growth of *Phytophthora erythroseptica* in nonsterilized soil can be determined with a simple plate procedure. The surface of nonsterile soil in petri dishes is made smooth and inoculated centrally with a 6-mm mycelial disc of the fungus. After incubation, the extent of linear growth is measured. This direct microscopic reading is verified by taking with sterile forceps samples of the central inoculum and of soil 2, 6, and 10 mm distant from the inoculum disc, and introducing these into wounds of healthy susceptible potato tubers. Development of pink rot indicates presence of the fungus. Pieces of lesions (6 mm diam) from the infected plants may be placed on a soil surface and tested in a similar manner. Thus, by relating extent of growth to size and kind of food base, the survival potential of a pathogen in natural soil may be estimated.

4. Cellulose-film bags (Stover, 1958b).

Cellulose bags may be used to determine effect of diffusible substances in soil on spore germination and growth of fungi such as *Fusarium oxysporum* f. *cubense*. The bags are prepared by adding field soil to washed dialyzing tubes (2.8 × 15.0 cm). The soil is saturated with distilled water so that the soil solution is in continuous contact with the cellulose film. The bags are tied to a rack, openings punched in the top for gas exchange, and then thoroughly washed with a stream of sterile distilled water under pressure. They are dipped into a warm 1.5% agar medium containing spores of the test fungus, and then suspended in a moist chamber. After 24-72 hr the bags are cut open, the soil is washed from the inner side, and the film is spread out in petri dishes for microscopic examination. Germination of spores and extent of growth can be observed and related to treatment of the soil.

B. SURVIVAL

Studies of survival of microorganisms in soil usually include adding the organism to soil, incubating the infested soil under defined environmental conditions for a period of time, and, finally, recovering the organism from the soil. Recovery is usually the difficult part, and may require special isolation procedures suitable for the particular organism. Many of the isolation procedures described in Chapter 5 are used to recover specific pathogens from soil. Descriptions of some additional methods follow.

1. Assessing disease severity.

This involves relating survival of an organism in soil to disease severity. This subject is adequately covered in this Chapter in section E, Inoculum Density and Potential. A typical procedure of this type is that used by Schreiber and Green (1962) to study survival of *Verticillium albo-atrum*. Fungal mats grown on Czapek's agar (4-6 days for mycelial-conidial inoculum, or 4 weeks for microsclerotial inoculum) are homogenized in water and mixed with vermiculite to facilitate uniform distribution in soil. Inoculum concentration, temperature, moisture, or other incubation factors may be studied. Samples of the soil are taken at intervals and seeded with susceptible indicator plants (tomato). Percentage of infected indicator plants along with a disease severity index is related to survival.

2. Addition and recovery from soil by special marker techniques.

Corynebacterium insidiosum, which causes a wilt of alfalfa, grows slowly and is usually overrun or inhibited by other organisms in dilution plates. This difficulty is eliminated with a special procedure developed by Nelson and Semeniuk (1963). The bacterium is grown on a modification of Burkholder's agar (Burkholder, 1938): approximately 300 g of sliced potatoes are boiled and the broth is made up of 1 liter. To this are added peptone, 5 g; K_2HPO_4, 2 g; sodium citrate, 1 g; asparagine, 0.6 g; dextrose, 6 g; and agar, 12 g.

A turbid cell suspension and requisite amounts of water are added to 100-g replicates of previously air-dried soil in 300-ml flasks. Concentration of cells in the suspension is determined with a hemacytometer. Where an experiment requires testing under low moisture, the bacteria in the suspension are coated and air-dried on sterilized silica sand and the latter is handmixed with soil of adjusted moisture. At desired sampling times, 20 g of soil are removed from a flask, shaken 10 min in 180 ml of water and serially diluted. The end dilutions are high so that the number of indigenous soil bacteria is reduced substantially. One-ml amounts of the final dilutions are placed in petri dishes and mixed with modified Burkholder's agar containing 250 ppm of Acti-dione to retard fungal development. After 7-14 days at room temperature (20-25 C), blue to purplish colonies of *C. insidiosum* are discernible and readily differentiated from greenish blue colonies of gram-negative bacteria. Colony counts/g dry soil at incubation intervals are compared with numbers of cells added at the beginning of the experiment.

Worf and Hagedorn (1961) devised a method for following the relative development of two species of *Fusarium* that attack peas, *F. oxysporum* f. *pisi* and *F. solani* f. *pisi*. Sand-loam mixtures, sterilized 1-3 weeks previously, are seeded with inocula of both fungi. On appropriate dates, cores of soil are taken with glass cylinders calibrated to hold 1 ml. Adhering soil particles are cleaned from each cylinder and the soil column is forced out with a rod into 100 ml of sterile distilled water in a 250-ml flask. The flasks with soil suspensions are agitated on a mechanical shaker for 15 min and serial dilutions prepared. One-ml portions of the desired end dilution are plated on PDA buffered at pH 3.1 with phosphate buffer and maintained at 20 C. Under these conditions growth of many fungi and all bacteria was inhibited, and colors distinguishing colonies of *F. oxysporum* (purple) and *F. solani* (white) developed within 5-7 days. With this isolation procedure, influence of soil treatments on survival and inoculum density can be studied.

3. Burial and recovery of artificially infested organic materials.

The method devised by Garrett (1938) is often used to study survival of fungi in soil. Pieces of wheat straw (or other organic materials) are cut into about 2.5-cm lengths with nodes included. They are boiled for 3 hr in a string-bean decotion and left in it overnight. They are then autoclaved in flasks in lots of about 1000 and inoculated with 300 g sand plus 3% cornmeal containing a culture of the organism to be tested. The cornmeal-sand inoculum is shaken with the straws to obtain simultaneous uniform inoculation. The flasks are then incubated for an appropriate period; the straws are removed, washed free of adhering sand and cornmeal, and placed in lots of 50 into envelopes until used. Each lot of 50 straws is buried in a tumbler of soil. Environmental conditions are adjusted as desired. To determine viability, straws are recovered at intervals, washed free of soil, and used to inoculate susceptible plants. Presence of the pathogen is determined by symptom development.

4. Burial and recovery of inoculum attached to inert substances.

a. **Fiberglass.** This material in the form of tape or threads can be employed for survival studies (Legge, 1952; Caldwell, 1958). Appropriate lengths of sterilized fiberglass tape (2.5 cm wide and 75 μ thick) or fiberglass threads (75 μ thick) are placed on an agar medium in petri dishes. The agar is inoculated with the fungus to be studied. When fungal hyphae have become interwoven with the threads and produce the desired reproductive structures, the tape or threads are lifted from the agar and placed in soil. Noninoculated controls of sterilized glass tapes or threads should be buried also. Periodically, the tapes or threads are removed, washed, stained in phenolic rose-bengal or other suitable stain, and examined directly with a microscope. To determine viability of the fungi interspersed in the individual threads, small pieces of unstained threads are placed on agar media in petri dishes, or in nutrient solutions in flasks.

b. **Nylon.** Nylon mesh is a suitable material on which fungal hyphae can be attached and buried in soil (Old and Nicolson, 1962; Mixon and Curl, 1967). Either monofilament mesh (400 squares/cm^2) or multifilament (1300 squares/cm^2) can be used. Autoclaved squares (1.5 cm) of nylon cloth are placed on the surface of nutrient agar inoculated with the test fungus. When the fungus has grown across the nylon, the squares are stripped off and buried in soil. Soil treatments are varied according to the information desired. After a certain period, or at intervals, the cloth is recovered and soil particles are washed off by shaking gently in water. The squares are then stained with cotton blue and mounted in lactophenol on glass slides. The small squares that comprise the mesh are used as standard fields for rating lysis under a microscope.

Waid and Woodman (1957) used a similar method for trapping out natural mycelia in soil. Therefore, a proper control (buried nylon squares without the test fungus) should be included. It would be helpful if the test fungal mycelia could be easily distinguished from naturally occurring mycelia from soil. A detailed procedure for counting should be recorded, since confusion might result from observation of branching mycelia and aggregates characteristically formed by certain Basidiomycetes.

Studies on survival of sclerotia in soil can be facilitated by using nylon bags (Williams and Western, 1965). Sclerotia are placed with a small quantity of sieved soil in 7.6-cm-square nylon bags, and buried at various depths in soil. At intervals, sample bags are removed, and the sclerotia are washed and subjected to conditions suitable for germination.

C. SOIL FUNGISTASIS

An inhibiting effect on germination of certain fungal spores has been demonstrated in a wide variety of soils. Though most investigations have involved germination of spores, the phenomenon of soil fungistasis may also be applied to inhibition of growth of hyphae in soil. Soil fungistasis implies inhibition only, and should not be confused with lysis of hyphae that often occurs following germination of spores. The following are examples of methods that demonstrate soil fungistasis.

1. Use of cellophane (Dobbs and Hinson, 1953).

Cellulose film, about 20 μ thick, is cut into 5-cm squares, boiled to remove coating material, and lightly autoclaved. While still damp the squares are dusted or sprayed with spores of the test fungus. They are folded with the spores inside, partly buried in moist soil, and pressed firmly to secure intimate contact of the film with the soil surface. Periodically, the squares are removed and examined for germination of spores and growth of the test fungus on the parts of the film in contact with soil. Results may be obscured after a few weeks by bacterial, insect, or mite attacks on the film. If the sweet dressing commonly on commercial cellulose film is not removed completely by boiling in water, bacterial growth may occur within a few days.

2. Agar discs (Jackson, R. M., 1958).

Agar discs are prepared by pouring melted 2% water agar into flat-bottomed petri dishes placed on a level surface, so that a layer of agar just under 1.5 mm thick is obtained. Discs are cut from the agar with a flamed cork borer, 7.5 mm in diam, and removed with a flamed scalpel. Soil samples are sieved, mixed, and placed in petri dishes. Moisture content, pH, addition of supplements, and degree of sterilization may be experimental factors. Squares of sterile Whatman No. 1 filter paper (4 cm) are placed on the surface of the soil in each petri dish, and an agar disc is placed on each of the squares. Either immediatley, or after 1-4 hr of incubation to permit diffusion of substances from soil into the agar, the surface of each disc is inoculated with a drop of a spore suspension of the test fungus in distilled water. If control and treated discs are inoculated with samples of the same spore suspension, standardization of the spore suspension is not necessary. Controls are similarly prepared, but the agar discs are placed on filter papers saturated with distilled water in dishes containing no soil. The petri dishes with lids replaced are incubated at selected temperatures for the minimum period required to obtain good germination of the spores on the control discs. After incubation, the discs are removed to microscope slides and covered with lactophenol and a coverslip. Numbers of germinated and ungerminated spores in random microscope fields on each disc are recorded.

The method was modified in our laboratory to determine soil fungistatic effects on growth of mycelia. An 11-cm circle of sterilized filter paper is placed on the surface of a layer of moistened soil in a petri dish. Excess paper is folded around the edge of the dish. A layer of warm, sterile, 2% water agar is poured onto the filter paper. Dishes thus prepared are incubated 24 hr at 5 C for diffusion of substances from the soil into the agar. Then, a mycelial disc (3-5 mm) cut from an actively growing culture in a petri dish is placed on the agar surface. After incubation at an appropriate temperature, growth of hyphae is measured and compared to that on a similar layer of agar in a dish without soil. The technique is suitable for fungi such as *Rhizoctonia*, *Pythium*, and *Fusarium*.

3. Agar slide (Chinn, 1953).

Fungi or actinomycetes to be tested are grown on PDA or nutrient agar slants. After incubation, the cultures are suspended in 10-ml samples of a detergent, "Tween 20" (diluted 1:5000). The suspensions are passed through sterile, fine copper screens and transferred into 250-ml beakers, each containing 200 ml of sterile 1% agar which has been previously melted and cooled. The soils to be used are sieved, moistened to about 50% of their moisture holding capacity, and placed in suitable containers. Clean, sterile microscope slides are dipped into the agar suspension of spores and inserted vertically 1 cm into the soil; several slides are placed in each container and covered with additional soil. The containers are covered to conserve moisture during incubation. At intervals, slides are lifted with as little disturbance as possible, rinsed with water, and the agar on one side is removed. The slide is rinsed again in water and stained by immersion in a dilute solution of cotton blue in lactophenol for 10 min. When the slides are examined with a microscope, data may be taken on spore germination, length of germ tubes and hyphae, and lysis of germ tubes and hyphae.

4. Membrane filters (Adams, 1967).

Moisture of the test soil is adjusted as desired and the soil placed in Coors porcelain crucibles (size 0, high form); these are placed in moist chambers to prevent drying. A spore suspension of the test fungus is washed by filtration, resuspended, and diluted to about 10^5 spores/ml. One ml of spore suspension is added to a 25-mm Type HA Millipore membrane filter mounted over a filter holder, and the liquid is removed by vacuum filtration leaving approximately 10^5 spores on each filter. The filters with spores are buried in a slit in the soil and the soil is pressed against the filter. The crucibles of soil are then incubated in the moist chamber. Membrane filters with a grid marked on one surface are desirable to permit identification of the surface on which spores are placed.

The filters are gently removed from the soil later, placed on microscope slides, and covered with lactophenol trypan blue (liquid phenol, 11.2 ml; glycerin, 10 ml; lactic acid, 10 ml; distilled water, 8.8 ml; trypan blue, 0.02 g). The slides are then steamed for 3 min over a water bath at 70 C, and the stained filters are placed on the lower half of a filter funnel and washed by vacuum filtration with clean lactophenol. They are washed again with glycerin and mounted in glycerin on glass slides for spore germination counts.

5. Direct method (Lingappa and Lockwood, 1963).

This involves placing spores directly upon the smoothed surface of soil in a petri dish, subsequently staining the spores *in situ,* then later recovering a sample of the spores with collodion film or agar. A loam soil with 25% moisture is ideal for preparing soil surfaces for assay. Twenty g of soil are compressed in a small petri dish (50 X 15 mm), and the surface is smoothed with a bent spatula blade; the final soil mass should be about 4-5 mm thick. A flat, compact, smooth surface assures uniform distribution of spores, good contact with the soil surface, and minimal removal of soil particles when spores are recovered. Drops of a spore suspension (concentration properly adjusted) are applied to the soil surface. Spores are similarly placed on water agar for reference. After incubation, 2-3 drops of aqueous phenolic rose bengal solution (1% rose bengal, 5% phenol, and 0.01% $CaCl_2$) are applied to the soil surface and allowed to diffuse into the soil. This stains and kills the spores and germ tubes but does not stain soil particles. Stain solutions containing oils or glycerol may prevent adhesion of spores to recovery media. Drops of collodion [1.5% pyroxylin (Parlodion) in 1:1 (v:v) absolute ethanol and ethyl ether] are placed on the soil surface and allowed to spread and dry to a thin film. The film is removed with forceps and placed in a drop of mineral oil (medical) on a glass slide and covered with a coverslip for microscopic examination.

If sterilized soil, inoculated with specific microorganisms, is used care must be taken to avoid contamination. Plastics other than collodion may be used for spore recovery: 1-2% polystyrene (Falcon Plastics petri dish) in benzene-toluene (2:1); or Epolene C-10, a polyolefin resin. Also 3% water agar, applied to the soil surface, may be advantageous to recover spores for viability test before staining. The method could be applied to other studies of the fate of fungal propagules and mycelia on soils. However, it may not be suitable for all species, since percentage recovery and staining of spores will vary.

6. Quantitative assay (Chacko and Lockwood, 1966).

The direct assay of soils for fungistasis, described by Lingappa and Lockwood (1963) eliminated certain disadvantages of employing substrates (agar discs, cellophane, etc.) which introduced undesired nutrients or physical factors. But this and previous methods were not considered quantitative. The following is a modification of the direct soil surface method for quantitative studies. Natural (nonsterilized) soil is serially diluted with soil sterilized with ethylene oxide vapors (in a closed chamber for 24 hr). Moisture is adjusted to 25%, and 20 g of soil of each dilution are placed in petri dishes (50 × 15 mm), compacted lightly, and smoothed. Aqueous suspensions of washed spores of test fungi are applied to the soil, incubated, stained, and recovered as outlined above.

To provide a uniform basis for comparison, germination of spores on various mixtures of natural and sterilized soil is expressed as the percentage of that occurring on sterilized soil. Results are plotted on log-probability paper with the ratio of sterile: natural soil on the log axis and the germination percentage on the probit axis. Resulting graphs are straight lines. At any given percentage germination of the test fungus, ratios of sterile: natural soil of test soils can be compared to indicate relative quantity of fungistasis. A soil requiring a low sterile: natural soil ratio to induce 50% germination (the point chosen for comparison) will be comparatively less fungistatic than a soil requiring a high ratio to induce the same percentage germination.

D. COMPETITIVE SAPROPHYTIC ABILITY (CSA)

Garrett (1956a) introduced the expression "competitive saprophytic ability" to distinguish between *soil-inhabiting* pathogens with a high CSA and *root-inhabiting* pathogens with low CSA. Success of an organism to colonize organic substrates in competition with other organisms is determined by certain physiological characteristics: (1) rapid growth rate, abundant sporulation, and rapid germination of spores; (2) necessary enzyme system for breaking down substrate tissues; (3) production of antibiotic substances; and (4) tolerance to antibiotics produced by other microorganisms. The significance of this phenomenon to plant disease incidence and control lies in its relation to survival time of soil-borne pathogens in natural field soil.

The competitive saprophytic ability of soil fungi may be tested by a method described by Butler (1953) and further developed by Lucas (1955). It is sometimes referred to as the "Cambridge Method," and was initially used to study saprophytic colonization of wheat straw by the cereal root-rot fungi, *Helminthosporium sativum, Curvularia ramosa, Ophiobolus graminis* and *Fusarium culmorum,* under natural soil conditions. Test pieces of straw are buried in a graded dilution series of maizemeal-sand culture of the test fungus (diluted with nonsterilized soil). The maizemeal-sand medium contains 100 parts sand, 3 parts ground maizemeal, and 13 parts water (by wt). After sterilization, each half-liter flask of medium is inoculated with the test fungus and incubated for the desired time. Pieces of straw, selected for uniformity, are cut and different lots treated with water, dextrose solution, sodium nitrate, or other materials according to test fungus and purpose of the experiment. Thorough wetting is ensured by soaking under reduced pressure for 2 hr. Straws may then be autoclaved or left unsterilized. Soil to be mixed with inoculum is screened and adjusted to about 45% FC. Inoculum is added to the soil in proportions of 100:0 (inoculum control), 98:2, 90:10, etc., the final mixture being 0:100 (soil control). Immediately after mixing, 50 straws per 200 g of inoculated soil are evenly distributed throughout the mixtures, which are then placed in glass jars previously adjusted to a constant weight with clean fine gravel. The inoculum control straws may be added to pure inoculum in 350-ml flasks. After incubation the straws are washed free of soil and inoculum, and saprophytic colonization is assessed by isolation on agar plates, or by Garrett's (1938) wheat-seedling test, or by incubation on moist sand to promote sporulation. The agar plate assay is considered least desirable of the three. In the wheat-seedling test a wheat seed is placed, germ end down, within the lumen of each piece of straw and planted in a seed-tray of sand. Seedlings are grown in a greenhouse for one month and rated for infection. In all assays, with proper correlations, data are expressed as percentage of straws colonized.

In a similar manner, organic materials other than wheat straw are often used to study saprophytic colonization in soil. Cellophane (Tribe, 1957), chitin (Okafor, 1966), banana leaf (Newcombe, 1960), buckwheat stems (Papavizas and Davey, 1959), and rhizomes of *Convolvulus sepium* (Matturi and Stenton, 1958) have been used successfully.

E. INOCULUM DENSITY AND POTENTIAL

"Inoculum potential" as defined by Garrett (1956a) refers to the energy of growth of a parasite available for infection of a host at the surface of the host organ to be infected. Since "inoculum potential" is a concept that has no measurable units, methods to determine its value depend on measurements of factors that affect it. Such factors include the density and the nutritional status of the inoculum. Measurements can also be made of physical and chemical environmental factors of the soil that affect inoculum survival, density, and nutrition of the inoculum.

Methods for increasing inoculum and infesting soil with pathogens are described in detail by Tuite (1969). Three typical methods are involved in measuring inoculum density: isolation from soil of the infecting units with quantitative methods such as the dilution plate, use of trapping methods to determine number of infecting units, or indexing the soil for its capacity to produce disease. Relating a specific concentration of inoculum under defined environmental conditions to a disease expression would constitute an estimate of inoculum potential. Since the literature is voluminous on this subject, the following descriptions of methods are selected ones which should give the reader a cross section of the different kinds of techniques used.

1. The most-probable-number method (Maloy and Alexander, 1958).

A spore suspension of the pathogen is prepared and mixed with sterile soil to give a known number of spores per unit weight. This infested soil is mixed with unsterilized soil, which does not contain this particular pathogen, in ratios of 1:1, 1:9, 1:99, etc., or in another convenient dilution series. Field soil, naturally infested with the pathogen, is diluted with sterile soil in a similar manner. The diluted samples are used to inoculate host plants; the particular method of inoculation depends on the type of pathogen and host. After incubation, presence or absence of disease symptoms is recorded. By use of a table of "most probable numbers" such as that given in Halvorson and Ziegler (1933), the most probable number of spores in the first dilution resulting in symptom development in all replicates can be obtained. Multiplication by the dilution factor gives an estimate of the number of spores per unit weight of soil.

2. Predicting disease severity.

Estimates of pea root rot severity in fields to be planted can be made by growing peas in samples of the field soil in pots in the greenhouse and by grading the roots according to an arbitrary visual disease index (Johnson, H. G., 1957). The samples are made up of sub-samples that represent the average topography of the field. This index must then be compared to the amount of root rot that occurs subsequently in fields from which the samples were obtained. It may then be useful for future predictions.

3. Serial-dilution endpoint.

A quantitative method developed by Tsao (1960) can be used to study distribution of *Phytophthora citrophthora* and *P. parasitica* in soil and estimate disease potentials *in situ*. Soils with different levels of *Phytophthora* infestation produce different end points of lemon fruit infection in soil serially diluted with sterile soil. From the endpoints, estimates of disease potential are then made. The dilution procedure is diagrammed in Fig. 10-2. Fifty ml of a pulverized soil sample are placed in a 500-ml wide-mouth, waxed-paper "split-cup" and divided into 2 equal portions. One half is labeled 1/1. Twenty-five cc of an autoclaved soil are added to the remaining half, mixed, and labeled 1/2 (1 in 2 dilution). This is repeated to give 7 dilutions up to 1/64. Deionized water is added to each cup and a ripe or near-ripe lemon fruit is placed in the cup with peduncle end up and the other end immersed in the soil-water mixture. The fruits are incubated at 25 C for 6 days and examined for brown lesions, which indicate a positive infection. The Disease Potential Index (DPI) is the reciprocal of the highest dilution that yields brown-rot lesions. For example, if the highest dilution allowing lesion formation is 1/32, the DPI = 32. Cups of sterile soil provide controls. Soils of known low infestation would not require extending dilution to 1/64. Finally, the DPI of *Phytophthora* may be related to degree of root damage.

4. Use of soil microbiological sampling tubes.

The following is a soil infestation procedure used by Martinson (1963) to study inoculum potential of *R. solani* with soil microbiological sampling tubes (see Chapter 2). Inoculum is obtained by growing the fungus for 30-45 days at 25 C in 2-liter Erlenmeyer flasks on a sterile 2% cornmeal-sand medium (980 g white quartz sand, 20 g white cornmeal, and 250 ml distilled water). Inoculum and soil weights are computed on an oven dry basis. Soil for an entire experiment may be blended in a cement mixer. Inoculum for each level of inoculum density is mixed with weighed portions of the soil, and soil moisture is adjusted to about 16% by spraying water into the blending soil. Infested soil is placed into glazed crocks (1200 g/crock) and a microbiological sampling tube is inserted to a depth of 10 cm into the soil in the center of each crock. Six holes 2.5 cm deep are made in the soil around the sampling tubes with a metal template, and 4 radish seeds are planted in each hole. The crocks are covered with nonwaterproofed cellophane and randomly placed in shaded soil temperature tanks. The cellophane is removed after the sampling tubes are recovered, and pre-emergence damping-off (failure of seedlings to emerge) is recorded daily. With this procedure it was demonstrated that frequency of isolation of *R. solani* with microbiological sampling tubes correlate closely with pre-emergence damping-off of radish, as influenced by factors that affect inoculum potential (inoculum density, temperature, and fungicide concentration).

5. Soil "infectivity index" (McKee and Boyd, 1952).

This is a biological method of assessing soil for dry-rot disease of potato caused by *Fusarium* spp. Soon after being removed from soil, tubers are dipped

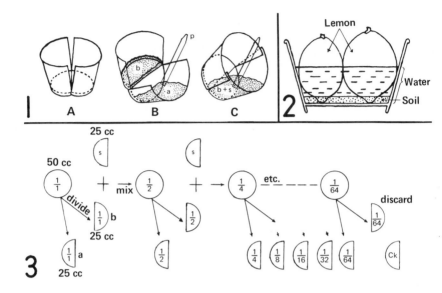

FIG. 10-2 (1-A). The "split-cup" made from a 500-ml paper cup by cutting the sides to form 2 connected, movable, equal halves. A different cup is used for each soil sample. (B,C) The split-cup shown in positions as a soil divider and a soil mixer, respectively. The 50 cc of soil after having been leveled at the cup bottom are divided into 2 equal halves as shown in B. One-half of the soil (a) is removed with the aid of a pot label stick (p) into a given dilution cup for the lemon trap test. To the remaining half (b) is then added 25 cc of sterile soil (s) and mixed thoroughly, as shown in C, to provide 50 cc of soil (b+s) of the next higher dilution. The mixing is done in the split-cup with a pot label stick by a circular motion while the cup, with portions of the movable halves overlapping, is held at an angle. Efforts are made by scraping with the pot label to avoid the adherence of wet soil particles on the cup walls. (2) The lemon trap test with each dilution cup, showing the positions of 2 lemon fruits, 150 cc of water, and 25 cc of soil in the cup. (3) Schematic diagram of the procedures involved in the dividing and the mixing of the soil in the serial dilution end-point method. The 8 semicircles in the bottom row, each representing 25 cc of soil, constitute a single replicate dilution series. (Tsao, 1960. Phytopathology 50:717-724.)

in a 2% solution of formalin for 30 sec to prevent natural infection through wounds caused during harvest. Cores of soil are collected from naturally infested fields to a 10-cm depth, and those from each field are bulked, mixed, screened, and small samples are stored in cotton-plugged tubes. Inoculations are made with a glass inoculator which has a rounded tip and flange at one end and is spoonlike at the other end. The rounded end is pressed into a tuber until stopped by the flange, then test soil is picked up with the spoon end and pressed into the hole. One hundred inoculations per soil are usually made, 2 in each of 50 tubers. More than 2 per tuber may cause coalesence of lesions. Inoculated tubers are stored under uniform temperature and humidity along with appropriate checks for about 6 weeks, then cuts are made across lesions and number of infections is expressed as percentage of the total number of inoculations. This "infectivity index" is regarded as a measure of the inoculum potential. Isolations from the lesions are necessary to identify specific fungi involved.

6. *Verticillium* "infection index."

An infection index devised by Wilhelm (1950) yields quantitative information on vertical distribution of *Verticillium albo-atrum* in different soils at various depths. Infection of tomato plants is used as a measure of intensity of the infestation. Soil samples are taken from desired depths in fields known to be infested with *Verticillium;* enough soil is taken from each depth to fill a 20-liter can. In the greenhouse, four 20-cm sterilized clay pots are filled from each can, and 10 susceptible tomato seedlings are transplanted into each pot. At the end of a growing period of 6-8 weeks, cross sections of the stems near the crowns are cut from each plant and placed on agar medium containing bits of dried sterilized plant material as nutrients. Presence of *Verticillium* and the relative quantity found are indicated by the infection index of the soil sample. The infection index is the percentage of plants, planted in a unit volume of soil, from which *Verticillium* is isolated after a given growing period. Theoretically, the infection index is related to both inoculum potential and inoculum distribution.

CHAPTER 11
ROOT EXUDATES

Root exudates are substances released by roots and may affect growth and activity of soil organisms in the rhizosphere. Amino acids, sugars, and vitamins have been identified in the exudate of a wide variety of plants. Other substances reported include organic acids, nucleotides, flavonones, enzymes, hydrocyanic acid, glycosides, auxins, and saponins. Following are descriptions of methods for collecting root exudates, detecting specific substances in exudates, and determining the effect of exudate on microorganisms.

A. COLLECTION

In the rhizosphere, substances in root exudates are quickly utilized, or otherwise acted upon, by the native microflora. Leachings of natural rhizosphere soil contain only small quantities of root exudate, and its concentration is usually masked by metabolites from the rhizosphere microflora. For this reason, root exudate collected for analysis should be produced under sterile conditions. Whole plants can be grown in nutrient solutions or sand culture in commercially available sterile growth chambers. Many laboratories do not have these facilities, and other devices have been used which are often more suitable than the larger sterile growth chambers.

1. Petri dishes (Flentje et al., 1963).

Exudates can be collected from small seedlings, such as lettuce or radish, grown for a few days in petri dishes. Surface-sterilized seeds are plated on sterile, distilled water agar for germination. Uncontaminated seedlings are transferred to petri dishes containing 10 ml of sterile, distilled water (20-25 seedlings per dish). After 5 or more days in the dark at 25 C, the seedlings and liquid are transferred to a sterile beaker, the seedlings are rinsed with sterile water, and the rinse liquid is combined with the rest of the exudate. The material is dried by evaporation under vacuum at 40 C or below, dissolved in a small volume of water, and then passed through a bacterial membrane filter into a weighed sterile tube. The sterile exudate is redried under vacuum, and the dry weight is determined. Sterile water is then added to make a 10% (w/v) solution.

2. Large test tubes (Rovira, 1956).

Seeds are immersed for 5-15 min in 0.2% $HgCl_2$ containing a wetting agent (Tween 80). If small grains are used, glumes should be removed before treatment. The seeds are washed free of $HgCl_2$ with at least six washings in sterile water and germinated on a nutrient agar medium. After incubation contaminated seeds are discarded. The germinated seeds are transferred to sterile tubes, 20 X 3 cm in size, containing acid-washed quartz sand and a plant-nutrient solution (Modified Crone's solution, Chapter 16). The tubes are placed in racks designed to keep the sand and roots in darkness and the tops exposed to light. "Daylight" fluorescent tubes may be used to supply 12 hr illumination alternating with 12 hr darkness each day. Additions of fresh nutrient solution are made at weekly intervals, and tubes containing contaminated plants are discarded at these intervals. Care should be taken to prevent wetting of the leaves to minimize accumulation of "leaf exudates" in the nutrient solution. To collect root exudates, the plants are removed from the sand and rinsed in sterile, distilled water. The wash water is retained. The sand is washed several times with sterile, distilled water, and all of the sand and root washings are collected. This material is either filtered or centrifuged to remove root debris. Then it is concentrated under vacuum to give 1 ml of extract for every 10 plants.

3. Sand tower (Ayers and Thornton, 1968).

This unit (Fig. 11-1) consists of a column of washed, ignited, quartz sand, moistened with sterile nutrient solution and supported by a porcelain crucible and glass wool at the neck of the lower portion of the tower. The upper planting tube is made from an autoclavable, 15-ml-capacity, conical polypropylene centrifuge tube with a 3-mm hole in the bottom. The planting tube, filled with small granules of perlite, rests in a rubber stopper fitted into the top of the tower. A narrow glass tube (which serves as an opening for additions of fresh nutrient solution and for rinsing at harvest) enters the chamber by another hole in the stopper and is bent to bring the opening near the lower end of the planting tube. The upper end of the rinsing tube is sealed with a glass plug in a rubber sleeve. Gas inlet and outlet ports at lower and upper parts of the tower are connected to cotton-plugged glass tubes by rubber tubing. Before planting, the top of this planting tube is covered with a glass sleeve plugged with cotton. The plant vessel is sterilized as a unit by autoclaving at 121 C for 3 hr.

Seeds to be used are surface sterilized as follows. They are washed 1 min in Triton X-100 (Alkylphenoxypolyethoxyethanol) and soaked in 0.17% $AgNO_3$ for 30 min. They are freed of $AgNO_3$ by washing for 1 min with sterile 1% NaCl and rinsing 6 times in sterile tap water. They are germinated for 1-4 days on sterile water agar, then vigorously growing sterile seedlings are placed aseptically in the planting tubes containing moist perlite which serves to adsorb materials

FIG. 11-1. Culture vessels used for root exudate studies. Left, single plant in sand culture. Right, wheat seedlings in solution culture. (Ayers and Thornton, 1968. Plant and Soil 28:193-207.)

exuded from the seed coat as the root system grows down into the sand. When the upper part of the plant has grown several cm, the protective glass sleeve is removed; several layers of sterile, dry perlite are added to the top of the planting tube, and the plants are sealed off with a layer of silicone rubber compound (Silastic RTV 502, Dow Corning Corp). Filtered, moist air, or gas mixtures are passed through the sand columns at intervals, or continuously. The plants are grown under artificial light, 800 ft-c, for 16 hr per day at 23 C. During the growth period, the root systems are protected from light by aluminum foil covers.

At harvest, the roots are rinsed *in situ* by flushing with three 100-ml volumes of distilled water. The bulked rinsings are filtered, then reduced in volume in a rotary vacuum evaporator at 40 C to 50-fold concentration or greater. The extracts are analyzed immediately or stored frozen.

4. Solution culture (Ayers and Thornton, 1968).

This unit (Fig. 11-1) consists of a glass tube, 5 cm long by 3.5 cm wide; the lower part is covered with nylon netting, held in position with a rubber filter adapter cone. A layer of moist perlite is placed in the planting unit on the nylon mesh. The cone and planting unit is fitted in the neck of a 600-ml-capacity, wide-mouth Erlenmeyer flask containing half-strength Hoagland's solution; the bottom of the mesh is approximately 6 ml from the liquid surface. The unit is covered with a glass sleeve that rests on the upper part of the flask. Atmospheric change is permitted in the upper chamber with a beaker cover supported by small rubber spacers. The apparatus is autoclaved partly assembled.

Seeds are surface sterilized and germinated as described for the sand tower unit. Germinated seedlings (15 wheat or 5 pea) are placed in the sterile planting unit and covered with additional perlite. The plants are grown under artificial light, 800 ft-c, for 16 hr per day at 23 C. During the growth period, the root systems are protected from light by aluminum foil covers.

5. Perfusion apparatus.

An Audus type perfusion apparatus (Audus, 1946) can be used to collect root exudates (Buxton, 1960). See Chapter 8 for a description of the apparatus. The unit, without a bypass tube with flow resistance R_2, is assembled with the upper chamber about half-full of washed sand. Side tube S is elongated and contains a cotton wool filter. After the unit is autoclaved, sterile nutrient solution is added, and a sterile germinated seedling is placed on the sand in chamber P. Tube T is positioned so that the percolating solution flows over the roots which grow into the sand. Samples of the solution may be obtained at any time through side tube S. They may be evaporated *in vacuo* to 1/10th of their original volume, analyzed immediatley, or stored frozen.

The apparatus can be modified to collect root exudates from roots of large plants, such as banana (Buxton, 1962). Plants are grown in nonsterilized soil in large perforated opaque plastic bags. A hole, 5 cm square, is cut in the side of each bag to which is attached a smaller bag containing Vermiculite soaked in Hoagland's solution. The Vermiculite in the subsidiary bag is arranged to make contact with the soil in the main bag, and the joints between the two are sealed with waterproof tape. The subsidiary bag is set at a $45°$ angle to the main bag and left in position for two weeks or more until roots grow into it. The subsidiary bag is cut open and the new roots are shaken free of Vermiculite, washed with water, held for 60 sec in 50% ethanol, and then rinsed in three successive changes of sterile water.

Single roots are carefully inserted into the sand of the upper container of the perfusion apparatus by leading them through a hole in the rubber stopper (Fig. 11-2). The root is held in the stopper with a Dispo nonabsorbent resilient plastic plug (Scientific Products, Waltham, Mass.) and a groove in the hole of the stopper prevents damage to the root after sealing. The upper container is made airtight with waterproof tape overlaid with Plasticene modeling clay. The root connection between the rubber stopper and the main bag containing the plant is bound in moistened cotton wool and loosely wrapped in tape overlaid with aluminum foil.

Roots treated in this way, and still attached to the plant, continue to make rapid growth in the sand. The percolating solution may be sterile, distilled water, or a nutrient solution. Sterility of the solution should be determined periodically by plating samples on nutrient agar.

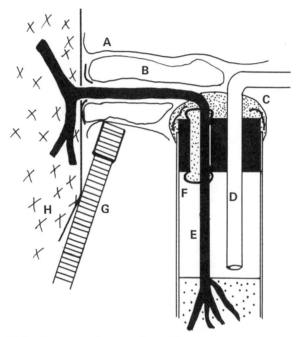

FIG. 11-2. Diagram of connection of banana root, previously grown in a subsidiary plastic bag, to upper vessel of root-perfusion apparatus. (A) aluminum foil covering; (B) cotton sheath around root connection; (C) modeling clay seal; (D) inlet tube for circulating water; (E) banana root inserted and growing into sterile washed sand; (F) Dispo resilient plastic plug; (G) edge of pot holding banana plant which is growing in soil in a plastic bag; (H) unsterile soil. (Buxton, 1962. Annals of Applied Biology 50:269-282.)

6. Absorption by filter paper (Schroth and Snyder, 1961).

A 11 X 25-cm sheet of Whatman No. 3MM filter paper is inserted between two 11 X 25-cm sheets of aluminum foil and autoclaved. Seeds (such as Pinto beans) are surface sterilized by treatment with 10% Clorox (sodium hypochlorite) for 3 min. The seeds are washed in sterile water and placed 4 cm apart near an edge of the filter paper. The paper is moistened with sterile distilled water and placed in a sterile 2.5 X 2.5 X 20-cm plastic container. Sterilized nonabsorbent cotton is used to press the soil snugly against the seeds so that the roots will grow flush with the paper. The container is inserted into a polyethylene bag, placed in the dark, and canted at a $60°$ angle to induce the roots to grow both downward and against the paper. Materials exuded from plant parts are absorbed by the filter paper. After an appropriate interval (5 days for Pinto beans), the germinated seeds are removed and tested for contamination by placing them in petri dishes containing potato-dextrose agar. To detect sites of exudation, the filter paper is dipped in appropriate reagents to detect the exudation of substances from the germinating seeds and roots.

B. DETECTION AND ANALYSIS OF SPECIFIC SUBSTANCES

A complete chemical analysis of root exudates is rarely attempted. The researcher is often concerned only with the presence or absence of different amino acids, sugars, phenolic compounds, or other substances that stimulate or inhibit root pathogens or rhizosphere microorganisms. Since he is often not concerned with absolute concentrations of substances in exudates, use of sophisticated gas chromatography techniques may not be necessary. For analytical techniques that could be used, the reader should refer to one or more of the many texts and laboratory manuals on thin layer and paper chromatography, spectrophotometric, and other techniques of analysis. Descriptions of procedures for root exudate analysis are found in the following publications: Katznelson et al. (1955), Andal et al. (1956), Rovira (1956, 1959), Rovira and Harris (1961), Buxton (1962), and Ayers and Thornton (1968).

C. EFFECT ON MICROORGANISMS

Many of the methods for determining the effect of root exudates on growth of microorganisms are similar to those used for other substances such as antibiotics, toxins, or growth regulators. If exudates are contained in a volume of sterile water or plant nutrient solution, it is often necessary to concentrate the solution *in vacuo* for positive results to be obtained. The concentrated, sterile exudate can be incorporated into nutrient liquid or agar media, or applied to paper discs on the surface of agar media. Such methods are described in detail in Chapter 13. In addition to this type of testing, solutions of known standard reagents, representing the substance or substances previously identified in the exudate, can be assayed for inhibitory or stimulatory effects on microorganisms. Various devices, or artificial rhizospheres, have been developed to determine the effect of root excretions on microorganisms. These devices also are useful for confirming the effect obtained by other methods. The following are descriptions of some assay techniques that illustrate the many approaches that can be used.

1. Spore germination (Buxton, 1962).

Root exudates, previously evaporated to 1/10 of their original volume, are dispensed in small test tubes, 0.2 ml in each. To this is added 1 ml of a medium permitting good germination of fungal spores to be tested. Spores of many strains of *Fusarium* germinate well in 0.5% sucrose in distilled water. A small quantity (0.2 ml) of a spore suspension of the test fungus is added. At intervals, percent germination of the spores is determined by the use of a hemacytometer. This is compared to percent germination in media without exudate.

2. Growth response (Rovira, 1956).

Sterile root exudate, the concentration of which should be determined by trial, is added to a sterile liquid medium in culture tubes. A bacterial, spore, or chopped mycelial suspension of the organism(s) to be tested is added. At intervals, growth responses are measured visually or by turbidity readings on a spectrophotometer.

3. Cellophane bioassay (Flentje et al., 1963).

This technique is useful for testing the effect of exudates on fungi (such as *Rhizoctonia* or *Pythium*) that have the ability to grow over the surface of water agar from a nutrient agar plug of the fungus. Concentric rings are cut in 2% water agar in 9-cm petri dishes (15 ml agar) with cork borers of internal diameters 7 and 12 mm. The agar between the cuts (for each concentric pair of rings) is removed leaving an "island" with a volume of approximately 0.1 ml separated from the rest of the agar by a "moat" 2.5 mm wide. With a micrometer syringe, 5-μl samples of the test material are added to sterile filter paper discs (6 mm) and one disc placed on top of each agar island. The agar in each petri dish is then covered with a cellophane membrane (8 cm). The membranes are cut from sheets of moisture-proof cellophane 20 μ thick, boiled in distilled water 30 min and autoclaved. A circular plug (7 mm) of inoculum is placed on top of the cellophane in the center of each dish. The technique is diagrammed in Fig. 11-3. The dishes are incubated until the membranes are covered with mycelium, then each membrane is lifted off and comparisons are made, before and after staining with lactophenol-cotton blue, between fungal growth over exudates and that over distilled water.

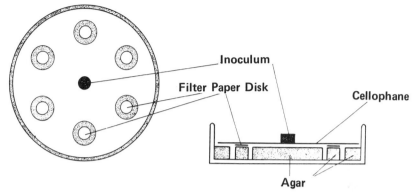

FIG. 11-3. Diagram of the technique used for the bioassay of root exudates on cellophane. (Flentje et al., 1963. Australian Journal of Biological Sciences 16:784-799.)

4. Germination of chlamydospores of *Fusarium*.

This direct method was developed by Schroth and Snyder (1961) and Cook and Snyder (1965) using *F. solani* f. *phaseoli*. Inoculum of the pathogen is increased in a sandy loam soil by adding conidia and incubating until dormant chlamydospores are produced. Bean seeds are then planted in plastic containers of the soil; the containers have removable sides to facilitate examination of soil near roots. The soil is examined for germinated chlamydospores 16 hr after planting, and at intervals thereafter. This is done by removing the soil around a root or seed onto a glass slide with 2-3 drops of water. The slurry is stained with 0.1% acid fuchsin in lactic acid to give a deep red stain to the spores. A coverslip is applied and the smears are examined under the high power of a compound microscope. Lysis of the hyphae is recorded by determining percentage of spores with germ tubes in seven counts, 100 spores per count in each incubation interval. As germ tubes lyse and disappear leaving the original chlamydospores, the percentage of spores with germ tubes decreases. Time at which the decrease begins during the observation period and the rate and extent to which it occurs are considered when comparing effects on host parts or nutrient amendments on lysis.

5. Artificial rhizospheres.

a. **Collodion membrane sacks (Timonin, 1941).** Plants are grown aseptically in glass jars containing modified Crone's nutrient solution. The plants are removed at the desired intervals and the solution filtered through a bacteriological filter. The solution is placed in collodion membrane sacks and these are placed in the soil. Materials in the solution may diffuse through the membrane into the surrounding soil creating an artificial rhizosphere. The collodion membranes should be about 2.5 cm long and 0.6 cm in diameter and allow diffusion of not more than 0.02 ml of solution per hr. When the membranes have been in the soil for several days, they are removed and the numbers of microoganisms in the rhizosphere soil determined. Special types stimulated by the diffused material may be identified. A comparison with control soil (collodion sack containing sterile, distilled water) should be made. With this method, the effects of exudates from disease-resistant plants upon the soil microflora may be compared with those from susceptible plants.

    b. **Cellophane bags (Kerr, 1956).** Cellophane sheets, 25 μ thick, are boiled in water to remove coating material, then made into bags by using adhesive Cellophane to seal the sheets together. Seeds are placed in the bags and partially buried in soil in glass jars; the soil may be previously sterilized and inoculated with the microorganism under investigation. After incubation for various periods, the bags are removed from the soil and washed to remove adhering soil. The plants are examined for stunting and necrosis of root material. Such conditions indicate that the organism produces a toxin that is cellophane-diffusable. Aggregation of the organism on the outside of the bag indicates that the plant produces a cellophane-diffusable material which stimulates development of the microorganism under study.

CHAPTER 12
SCREENING SOIL MICROORGANISMS FOR ANTAGONISM

The first step of an exacting method for isolating antagonistic organisms from soil involves obtaining colonies that represent the total population on dilution plates or soil plates. Isolates are selected at random from these plates and transferred to pure culture. After incubation and growth they can be assayed individually for antagonism to the test organism by (1) one or more of the methods using agar media (Chapter 13), (2) one or more of the methods using sterile culture filtrates (Chapter 13), or (3) by determining their ability to parasitize or colonize the mycelia or sclerotia of the test fungus (section E, this chapter). Such individual testing of microorganisms for antagonism has distinct advantages, since each organism is assayed in a single petri dish, and an exact measure of its activity can be made. Growth rates and other physiological aspects of both the test organism and the potential antagonist can be taken into account, with the result that appropriate agar media and specialized techniques can be utilized. When several organisms are screened in a single petri dish, some antagonists might be missed due to overcrowding of colonies or failure of certain antagonists to grow or produce antibiotics on the isolation medium chosen. But testing large numbers of organisms individually is laborious and time consuming, and unless special information is desired, the screening techniques described below are adequate for routine assays of soil samples.

As a result of the intensive search for antibiotic producing microorganisms, a variety of these screening techniques has been developed. Although many were designed originally to isolate organisms which produce antibiotics effective for controlling human and animal diseases, they have been used extensively to find organisms active against plant pathogens; the basic principles of antibiotic action and antagonism apply in both cases. Specialized methods for isolating fungal parasites of mycelium and sclerotia of phytopathogenic fungi have been developed. Included in this chapter also are procedures for isolating bacteriophages and actinophages from soil. See Chapter 16 for agar media used for isolating various groups of microorganisms.

A. THE BACTERIAL AGAR PLATE (WAKSMAN, 1945)

Sol dilutions from 1:100 to 1:1,000,000 are prepared and 1-ml portions are pipetted into sterile petri dishes. A melted and cooled agar medium (suitable for growth of both test organism and antagonistic population), which contains bacterial cells, conidia, or fragmented mycelia of the test organism, is added to the soil dilutions. The dishes are rotated gently to disperse the cells in the medium, then incubated for appropriate periods before examination.

Bacterial suspensions should be prepared previously by adding 5-10-ml portions of a 48-hr broth culture of the bacterium to the agar medium which has been cooled to 42 C (Patrick, 1954). If conidial suspensions are to be used, it is best to filter the water suspension through a thin layer of sterile cotton to remove hyphae. Mycelial suspensions of nonsporulating fungi may be made by chopping cultures in a Waring Blendor. Mycelial suspensions of certain fungi, such as *Pyrenochaeta terrestris,* can be made by growing cultures in liquid media containing small particles of broken glass. The cultures are shaken at intervals so that the sharp edges of the glass chips cut the mycelium into sections (Freeman and Tims, 1955).

Agar media used will vary according to individual preference, and according to the type of antagonist desired. The medium should support the development of adequate numbers of the type of antagonist desired·and, at the same time, be favorable for the growth of the test organism. After an appropriate incubation period, dishes containing the soil dilution and test organism are examined. The number of organisms producing zones of inhibition is determined, and individual antagonists are transferred to pure culture. Antagonists may be retested by one or more of the methods described in Chapter 13.

Carter and Lockwood (1957) modified this method to detect soil microorganisms which lyse fungal mycelium. An agar medium containing conidia of a suitable fungus is placed in petri dishes. After the appearance of abundant mycelium, usually in 2-3 days, the agar surface in each dish is sprayed with approximately 0.6 ml of a 1:1,000 and/or 1:10,000 soil suspension. Lysis of mycelium is indicated by clear zones around colonies. Transfers to pure culture may be made for further study.

B. LAYERS OF AGAR

Large populations of microorganisms can be tested for antagonism by using layers of agar. The technique may be so designed that a population of microorganisms growing on a certain medium can be tested for antagonism against a sensitive microorganism requiring another medium. The agar layers, in the order added to the dish, are as follows (Kelner, 1948):
1. Foundation layer: a layer of about 15 ml of sterile agar medium, of a composition suitable for the growth of the antagonist(s).
2. Seeding layer: a layer of 0.5-1.0 ml of soft agar medium (0.25% agar), containing the diluted microflora to be tested.

3. Diffusion or barrier layer: 5-10 ml of sterile agar of special composition. In many cases this barrier is not necessary.
4. Test layer: 3 or more ml of agar medium (0.5-1.5% agar) containing a suspension of the microorganism, the inhibition of which is to be tested.

In testing soil for bacterial antagonists, the soil suspension should be so diluted that 30-50 colonies develop on the seeding layer. The agar in each layer should be left to harden before adding the next layer. Temperature of the media should be stabilized to 42-45 C in a water bath before adding dilutions of microorganisms. The dishes are incubated and microorganisms that are surrounded by clear zones of inhibition may be transferred for further study.

Freeman and Tims (1955) tested soil actinomycetes for antagonism to *Pyrenochaeta terrestris*. In their modification, the soil dilution is distributed in the foundation layer (Conn's glycerol-asparaginate agar). After 4 days of incubation, a seed layer consisting of a mycelial suspension of *P. terrestris* in Czapek's sucrose-nitrate agar is added.

Herr (1959) screened soil actinomycetes for antagonism to *Fusarium roseum, Rhizoctonia solani,* and *Verticillium albo-atrum*. His basal agar layer consists of 10 ml of 1.5% water agar. The second layer consists of 5 ml of 1% water agar containing the soil suspension. The dilution used should be determined by trial. Dishes containing the two layers are incubated at 25 C for 2 days, at which time actinomycete colonies are barely visible. Three-day-old cultures of the assay fungus cultured at 25 C in Czapek's sucrose-nitrate solution are macerated in a Waring Micro-blendor and mixed with an equal volume of Czapek's agar (2% agar) kept liquid at 47 C. Five ml of the fungus suspension are added as the third layer in each plate. The cultures are incubated at 25 C for 4-7 days for inhibition zones to develop.

Koike (1967) described a method for use with *Pythium graminicola*, a fungus which rarely sporulates in culture, but will grow over the surface of water agar in petri dishes. Cultures of *Pythium* are prepared and incubated for 5 days on water agar (2%) in petri dishes of a slightly larger diameter (90 mm) than those used for soil-dilution plates (87 mm). After a 5-day period of incubation of the soil-dilution plates (medium contains 2.5% agar), the agar in each dish of fungal colonies is transferred in its entirety onto a culture of *Pythium*. Before this, however, 5 ml of melted and cooled potato-dextrose agar (PDA) are spread over the surface of the *Pythium* culture. The soil-dilution agar disc is then inverted directly on the very thin PDA layer. Added nutrients in PDA stimulate growth of *Pythium* and result in a thick mycelial mat; inhibition zones around the colonies are more distinct. Dilution plates containing colonies of actinomycetes and bacteria are inverted directly onto the surface of *Pythium* cultures without an added layer of PDA. The nutrients in the isolation medium (sodium albuminate was used by Koike) stimulate growth of *Pythium* sufficiently. Zones of inhibition may appear after 24-48 hr.

C. SPRAY TECHNIQUES

Plates containing microorganisms (soil-dilution plates, etc.) are sprayed with a bacterial, conidial, or mycelial suspension of the test organism (Fig. 12-1). An ordinary spray atomizer is used commonly. The agar medium in the plate must support adequate growth of the antagonists and the test organism. Organisms producing zones of inhibition after suitable incubation periods may be transferred for further study. Stessel et al. (1953) screened 4 test organisms simultaneously with a spray technique. Soil-dilution plates, 7-14 days old, were sprayed with standard spore suspensions containing nearly equal concentrations of conidia of the 4 plant pathogens, each exhibiting no cross antagonism. Antagonists which produced large clear zones free from growth of any of the four pathogens were transferred for further study.

A spray apparatus (Stansly, 1947) is useful when refinement of the spray technique is desired, or when airborne spores become a problem in maintaining aseptic conditions. The apparatus (Fig. 12-2) consists of a de Vilbiss medicinal atomizer (No. 153 or 154) connected by friction to a compressed-air cut-off assembly (No. 633 or 648) which has a convenient trigger control. The assembly is connected to the compressed-air line through a reducing valve and gauge. The nozzle (G) is inserted into the lower end of the spray chamber (C) through a small rubber stopper and glass tube, which is then pushed through a large rubber stopper (E). A second glass tube (F) in the stopper, after passing through a solution of phenol or other disinfectant, is connected to a water aspirator. The spray chamber (C) consists of a pyrex tube 24 cm long, with an outside diameter of 9 cm. At the upper end of the chamber where the agar plate is held, a gasket (A) is constructed with 1.3-cm electrical tape and next to it a barrier (B) is made with layers of rubber tape covered with electrical tape. The gasket is constructed to permit a reasonably tight fit of petri dishes of normally varying diameter. The barrier permits the petri dishes to be held firmly during the spraying operation without danger of crushing the agar. Both gasket and barrier will stand repeated sterilization.

Aspiration serves a dual purpose. The reverse current produced prevents spray from leaking out during removal of a sprayed plate and its replacement with a fresh one. A second function of aspiration is to partly dry the sprayed plates before removal.

D. SOIL ENRICHMENT

The concept of soil enrichment involves adding to soil a simple or complex organic substance in order to stimulate development of organisms using this substance as a nutrient. Theoretically, under favorable conditions of temperature and moisture, one might be able to stimulate development in soil of protozoa by adding certain bacteria, nitrifying bacteria by adding an ammonium salt, cellulose-decomposing organisms by adding a cellulosic material, or bacterio-

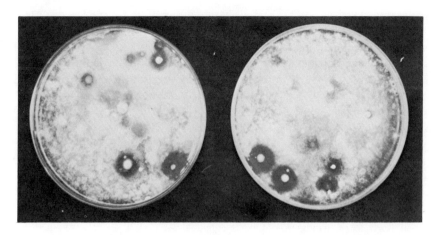

FIG. 12-1. Actinomycete-dilution plates sprayed with spore suspension of *Fusarium oxysporum* f. *tabaci*.

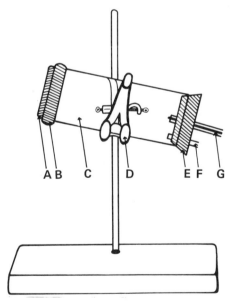

FIG. 12-2. Stansly's spray apparatus: (A) gasket; (B) barrier; (C) spray chamber; (D) clamp; (E) rubber stopper; (F) glass tube for aspirator connection; (G) atomizer nozzle. (Stansly, 1947. Journal of Bacteriology 54:443-445.)

phages by adding susceptible bacteria. Attempts by Waksman and Schatz (1946) to increase the number of antagonistic bacteria by enriching the soil with specific susceptible organisms had only limited success. They found that increases in the development of specific groups of organisms resulted probably from the ability of such organisms to attack and utilize the living or dead bacterial cells added, and was not related necessarily to their antibiotic-producing ability.

A method for increasing the actinomycete population (including antagonistic actinomycetes) was developed by Tsao et al. (1960). Using the spray technique of Stessel et al. (see above), they found that many of their plates were overrun with fast-growing fungi. In order to reduce fungi in the soil to be assayed, the soil was air dried and then incubated in a moist chamber with $CaCO_3$. Drying of the soil reduced the microbial population, and subsequent incubation of the soil with $CaCO_3$ changed the pH to favor growth of the remaining actinomycetes. With this method approximately 40,000 actinomycete colonies from 81 soils were examined. Two hundred fourteen colonies that produced antibiotics effective against 5 plant-pathogenic fungi were selected and isolated. A majority of these produced inhibition zones of 24 mm or more in diam. Details of the method follow.

Soil samples are pulverized and spread in thin layers in trays at room temperature for 5-9 days to dry, then stored in glass containers until used. One g of each sample is mixed with 0.1 g of powdered $CaCO_3$ and incubated in a closed inverted petri dish at 26 C for 7-9 days. High relative humidity is provided by saturating, with sterile distilled water, 8 pieces of filter paper fastened with arched wire in the bottom of the dish.

After incubation, dilutions for actinomycetes are made. Dilution plates are poured with soybean meal glucose agar adjusted to pH 7.9-8.1. Ingredients of the medium are listed in Chapter 16. After 5-6 days of incubation the agar surface of dilution plates, containing 10-20 actinomycete colonies each, are sprayed with water suspensions of the test organism. If desired, mixed conidial or cell suspensions of up to five mutually nonantagonistic test organisms capable of growing on the isolation medium may be sprayed on each plate. After incubating the sprayed plates for 3 days at 26 C, actinomycetes surrounded by clear zones are isolated.

E. PREDATION AND PARASITISM AMONG MICROORGANISMS

The practical significance of direct predation and parasitism among microorganisms is not well understood, but the phenomenon apparently is widespread in nature. Predation may consist of simple feeding of the microfauna on other living organisms, such as that exhibited by Protozoa ingesting bacteria, and by mites feeding on bacteria, fungal mycelia, and fungal spores. Predacious mites feed on nematodes and predacious nematodes feed on other nematodes.

Also, certain species of nematodes feed on fungi, among which are important plant parasitic species. A preference for root-rotting fungi has been demonstrated for the mycophagous nematode, *Aphelenchus avenae*. This nematode develops much better when fed *Pythium arrhenomanes, Rhizoctonia solani, Thielaviopsis basicola, Verticillium albo-atrum, Fusarium solani*, and *Armillaria mellea* than when common soil saprophytes are used as food sources (Rhoades and Linford, 1959; Mankau and Mankau, 1962).

More elaborate and sophisticated mechanisms of parasitism are exhibited by segments of the soil microflora. Among the best known are the predacious fungi that attack and capture nematodes. Special structures, such as sticky knobs, sticky spores, adhesive networks, and constricting rings, are produced that serve to attach the fungus to the nematode. Germ tubes arise from these structures and grow into the animal body, and eventually the body is filled with mycelium which destroys the nematode host.

Some fungi are parasitic; they attack and destroy various species of other fungi. Weindling (1932) described the parasitic action of *Trichoderma lignorum* on *Rhizoctonia solani, Phytophthora parasitica, Pythium* spp., *Rhizopus* spp., and *Sclerotium rolfsii*. Aerial hyphae of *T. lignorum* coil around the host fungus, or otherwise make close contact, causing disintegration of the cellular structure and a release of cell contents. Also, action at a distance apparently is caused by secretion of antibiotic substances. Among the several fungal parasites of Oomycetes are *Trinacrium subtilis* and *Dactylella spermatophaga*, which are parasitic on oospores of pythiaceous root-rotting fungi (Drechsler, 1938). Infection is accomplished by perforation of the oogonial and oospore walls, followed by development within each oospore of a branched, somewhat lobulate, and rather massive haustorium. *Rhizoctonia solani* attacks a number of fungi by penetrating their aerial hyphae and establishing an internal parasitic mycelium, by coiling around their aerial hyphae or by a combination of both (Butler, 1957) (Fig. 12-3). Conversely, mycelium of *R. solani* is penetrated and destroyed by *Penicillium vermiculatum* and *Trichoderma* sp. (Boosalis, 1956). Sclerotia of a number of plant pathogens are attacked and destroyed by saprophytic fungi (Makkonen and Pohjakallio, 1960).

The usual method for determining whether a microorganism is parasitic on another involves individual testing in pure culture, but some screening procedures have been developed. The techniques described below are useful for isolating and studying microbial parasites of plant disease fungi. Methods for isolating predacious fungi that attack plant parasitic nematodes are described in Chapter 2.

1. **Colonization and parasitism of sclerotia.**

A qualitative technique for isolating microorganisms that attack living sclerotia was developed by Ferguson (1953, 1957). A high percentage of the

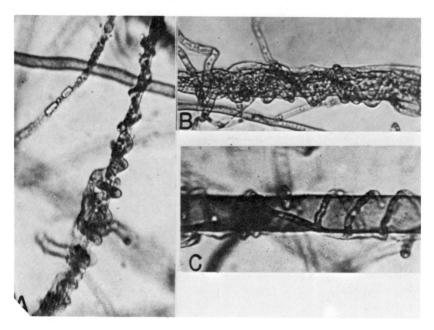

FIG. 12-3. *Rhizoctonia solani* hyphae coiling around and penetrating hyphae of fungi. A. *Pythium debaryanum;* B, C, sporangiophores of *Rhizopus nigricans.* (Butler, 1957. Mycologia 49:354-373.)

organisms isolated from sclerotia with this method were actively antagonistic to the mycelium. Organisms that were found colonizing viable sclerotia were species of *Trichoderma, Penicillium, Aspergillus, Streptomyces,* and *Plicaria fulva.*

Viable sclerotia are placed on moist soil in a petri dish or dipped in water dilutions of soil, and incubated at room temperature for several weeks. Specific incubation temperatures can be used to enhance colonization by various segments of the population. The sclerotia may or may not germinate during the incubation period. The sclerotia of *Sclerotium rolfsii* from *Delphinium* often germinated under the various conditions of incubation; germination of sclerotia of *Sclerotinia sclerotiorum* was less frequent. Colonization by microorganisms is based on the presence and development of organisms on the sclerotia. Pure cultures of the colonizers are obtained by picking off spores, mycelium, or bacterial colonies with a sterile needle, and transferring them to appropriate agar media.

Detection of natural infection of sclerotia of *Sclerotinia trifoliorum* by *Coniothyrium minitans* can be accomplished with a water-test method (Tribe,

1957). The sclerotia are placed in water in petri dishes for 1 hr and the region around the sclerotia is examined for patches of spores exuded from pycnidia of *C. minitans.* The water test underestimates numbers of infected sclerotia, but it is suitable for routine screening.

Extensive research on colonization and parasitism of sclerotia of many plant parasites by fungi and bacteria has been conducted in Finland by Makkonen and Pohjakallio (1960). At least 14 species of fungi and one bacterium were found to parasitize and damage sclerotia of one or more of five pathogens (Fig. 12-4). Contamination is effected by placing sclerotia for 2 hr in water in which mycelia and spores of the presumable parasites are suspended. The sclerotia are then transferred to the surface of autoclaved moist quartz sand in petri dishes. In order to maintain high humidity, the petri dishes with the inoculated sclerotia are enclosed in plastic bags and incubated at room temperature (approx. 20 C). Fungi which are able to destroy the sclerotia and which can be isolated from them are considered to be parasites.

Organisms which colonize sclerotia of *Sclerotium rolfsii* can be isolated with a method developed by Curl and Hansen (1964). Groups of 100 sclerotia are placed in 10 ml of sterile water in test tubes and shaken vigorously on a

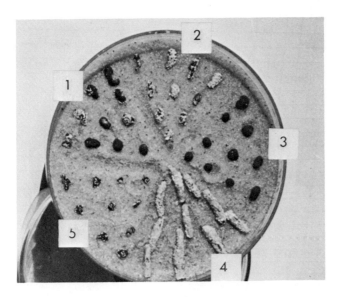

FIG. 12-4. Sclerotia contaminated with the fungus *Acrostalagmus roseus:* 1) *Sclerotinia trifoliorum,* 2) *S. borealis,* 3) *S. sclerotiorum,* 4) *Claviceps purpurea,* 5) *Botrytis cinerea.* (Makkonen and Pohjakallio, 1960. Acta Agr. Scand. 10:105-126.)

mechanical shaker for 30 min. The wash water is removed and this washing procedure is repeated twice. Finally, with a flamed and cooled glass rod, the sclerotia are thoroughly crushed to release organisms bound within, and the fragments are resuspended in 10 ml of sterile water. This is shaken as before and serial dilutions are prepared. One ml of the proper end dilution is transferred to each of several petri dishes and 10 ml of liquid agar media are added. Fungi may be isolated on Martin's peptone-dextrose agar with rose bengal and streptomycin (modified by increasing streptomycin to 100 μg /ml). Thornton's standardized medium is satisfactory for both bacteria and actinomycetes. Dilution plates are incubated at room temperature for 5-10 days and developing colonies are transferred to pure culture for further study.

2. Parasitism of fungal mycelium (Boosalis, 1956).

Penicillium vermiculatum and *Trichoderma* sp. parasitize the mycelia of *Rhizoctonia solani* (Fig. 12-5). Hyphae of the host fungus are invaded by penetration pegs developing from mycelium in contact with the host hyphae.

The host fungus is grown on a medium consisting of 100 parts sand, 1 part ground maize meal, and 13 parts water by weight. Nonsterilized soil is infested with the host fungus by mixing thoroughly 15-day-old inoculum at the rate of 800 g/23 kg of soil. Eight weeks later 0.5-g samples of the original inoculum, readily detected in the soil by presence of the sand, are removed. The bits of inoculum are rinsed in water until the mycelium is relatively free of soil and debris: The mycelium is examined microscopically. Fungi observed within hyphae of the host fungus can be isolated to pure culture by a dilution procedure. Single parasitized hyphae are picked from a dilution suspension under a dissecting microscope. Single hyphae obtained in this manner are transferred to 1.5% water agar (pH 4.8) and incubated for 3 weeks at 27 C. Hyphal tips are grown in pure culture and later retested for parasitism to the host fungus.

The "hot-water method" of isolation can be used if the parasite has a high thermal death point. (In this case, the thermal death point of *Penicillium vermiculatum* was 59.5 C.) The washed inoculum obtained from the soil is submerged for 1.5 min in water at 75 C, then plated out on 1.8% water agar of pH 4.8.' Isolates obtained are tested for parasitism in pure culture. The heat treatment apparently destroys soil-inhabiting fungi with thermal death points below 57-55 C. With this procedure it is possible to isolate parasites closely associated with unparasitized inoculum.

F. BACTERIOPHAGES

One of the few studies of bacteriophages that affect plant pathogenic bacteria is that of Crosse and Hingorani (1958) who described a method for isolating phages active against *Pseudomonas mors-prunorum,* the organism causing bacterial canker of stone fruits. Soils beneath diseased cherry, plum, and apricot trees were prolific sources of these phages. Cook and Quadling (1959) described a method for isolating from soil bacteriophages affecting *Xanthomonas phaseoli*. Their method is unique in that it is based on using mutant bacteria resistant to streptomycin. These methods may be used as guides for isolating bacteriophages that affect other plant pathogenic bacteria.

1. **Method for *Pseudomonas mors-prunorum* (Crosse and Hingorani, 1958).**

A composite 150-g soil sample is placed in a sterile glass jar and enriched with a 48-hr culture of *Ps. mors-prunorum* in nutrient broth plus 2% glycerine. The mixture is shaken periodically for 48 hr. The free liquid is decanted and partially clarified by filtration followed by centrifugation at 3,000-4,000 rpm for 30 min. Five ml of the supernatant liquid are transferred to a stoppered 20-ml bottle and shaken vigorously with 0.1 ml of chloroform to kill the bacteria. After the chloroform has settled out, 1 ml of the treated soil extract is withdrawn. Plaque formation is obtained by use of the poured plate method (Kleczkowska, 1945). This method consists of making successive tenfold dilutions from the original phage suspension in tubes containing 9 ml of 24-hr-old liquid cultures of the host bacterium. After shaking, 1-ml samples from appropriate dilutions are added to tubes containing 9 ml of melted agar medium

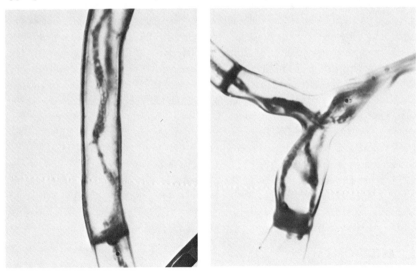

FIG. 12-5. Hyphae of *Penicillium vermiculatum* within hyphae of *Rhizoctonia solani*. (Boosalis, 1956. Phytopathology 46:473-478.)

cooled to 42 C. The agar medium containing bacteria and phages is shaken and poured into petri dishes. The agar medium used for *Ps. mors-prunorum* consists of 1% agar, 0.5% peptone, 0.3% yeast extract, and 2% glycerine. The medium is adjusted to pH 7.2 with a phosphate buffer.

After incubation for 24-48 hr, plaques varying in number from one to several hundred per plate may be observed. Single plaques are picked off, and the phages are propagated on young cultures of the bacterium in a broth of the same composition as the plating medium. The phages are purified by streaking a small drop of lysed culture across the dried surface of an agar plate previously sown with 1 ml of a 48-hr-old culture of the bacterium. Isolated plaques thus obtained are cut from the agar and repropagated in broth culture. The whole process is repeated 2 or 3 times and the phages are finally brought to high titer by serial transfers in appropriate bacterial broth cultures.

2. Streptomycin-resistant-strain technique (Cook and Quadling, 1959).

Mutants of the host bacterium (in this case, *Xanthomonas phaseoli*) resistant to 2,000 µg/ml of streptomycin are selected. About 50 g of soil are enriched with approximately 5×10^9 bacteria, suspended in 5 ml of liquid medium containing 500 µg/ml of streptomycin. This mixture of soil with host bacteria is incubated for 16 hr at room temperature. After incubation, 2 g of the enriched soil are suspended in 10 ml of liquid culture medium for 2 hr. Particles of soil and debris are then sedimented by low-speed centrifugation and the supernatant is tested for the presence of phage with the soft agar layer technique.

Soft agar layer plates (Adams, 1950) contain 2 layers of agar. The foundation layer consists of nutrient 1.5% agar with 500 µg/ml of streptomycin. The upper layer is prepared by cooling tubes of 0.7% agar to 46 C, each tube containing 2.5 ml of medium. The medium is held at this temperature in a water bath. One drop of a dense suspension of host bacteria is added to each tube. Samples (0.1-0.5 ml) of the supernatant containing the phage are added to the tubes; these are shaken and the contents are poured onto the surface of the hardened agar in the petri dishes. The bacteria grow as tiny subsurface colonies in this layer and are nourished by the nutrients in the foundation layer. Plaques appear as clear areas in the opaque layer of bacterial growth. Although antibiotic-resistant microorganisms other than the added host bacteria may be present, they should not interfere with plaque formation.

G. ACTINOPHAGES

Two methods are described below for isolating actinophages from soil. Both are suitable for isolating phages affecting plant parasitic actinomycetes such as *Streptomyces scabies* and *S. ipomoeae*.

1. **Shake culture method (Rautenshteyn and Kofanova, 1957).**

Samples of soil, 30 g each, are placed in 250-ml flasks containing 70 ml of meat-peptone bouillon. The flasks are incubated for 48 hr at 25-26 C on a shaker at 150-200 rpm. The contents are then filtered through Seitz filters and the filtrates are tested for presence of actinophages that cause lysis of actinomycete cultures.

The presence of actinophage is demonstrated on peptone-corn agar, consisting of (per liter): glucose, 10.0 g; NaCl, 5.0 g; $CaCl_2$ or $CaCO_3$, 0.5 g; peptone, 5.0 g; corn extract, 5.0 g; and agar, 20.0 g. The actinomycetes to be tested are sown on the surface of this medium in petri dishes. One to two drops of the filtrate are placed on the center of each freshly sown dish and it is tilted so that the filtrate is spread over a definite part of the agar surface. The cultures are incubated at 25-26 C for 24-72 hr. The presence of actinophage is indicated by complete absence of growth of the actinomycete on areas inoculated with filtrate or by smaller clear, lysed areas (plaques) within this zone. Complete absence of growth on the treated areas is probably due to a considerable amount of actinophage present in the filtrate. However, some soils possess antibiotic substances which may cause this condition to occur on the plates. Thus, subcultures should be made to dilute the active material. This can be accomplished by taking a small piece of agar from a sterile area and spreading it over the surface of a freshly sown plate with a spatula. After incubation, if phage is present a number of distinct plaques will appear on the agar surface.

2. **Soil-perfusion method (Robinson and Corke, 1959).**

An Audus-type soil-perfusion apparatus is used in this procedure (Chapter 8). In each unit a 50-g portion of 2-5 mm mesh, air-dried soil is perfused with 200 ml of distilled water at room temperature At daily intervals after 2 days of perfusion, 0.1-0.5-ml samples of perfusate are removed, filtered through a Millipore filter, and tested for plaque formation by the double layer soft agar technique.

The agar medium used is as follows (per liter): beef extract, 1.5 g; yeast extract, 2 g; peptone, 6 g; dextrose, 1 g; and agar, 15 g. The foundation layer consists of 12-15 ml of this medium in each petri dish. A 0.5-ml spore suspension of the actinomycete to be tested is mixed with 2 ml of the above medium (but with 0.75% agar) containing 0.1-0.5 ml of the filtered soil percolate. This mixture is poured over the hardened base layer. Plaques (Fig. 12-6) can be counted after 24-48 hr of incubation at 25 C.

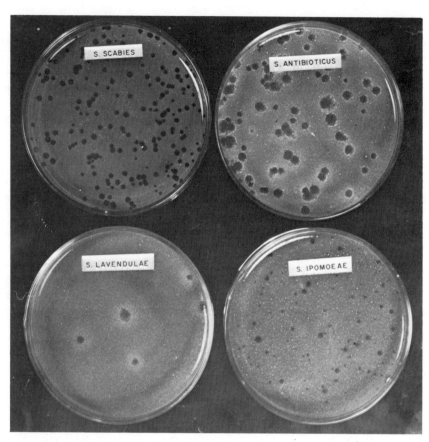

FIG. 12-6. Actinophage plaques. (Robinson and Corke, 1959. Reproduced by permission of the National Research Council of Canada from the Canadian Journal of Microbiology, 5, p. 482.)

CHAPTER 13
TESTING FOR ANTIBIOTIC ACTIVITY

Microorganisms obtained on soil-dilution plates, or with one of the screening methods previously described, may be tested individually for antagonism to specific organisms in pure culture. The information obtained indicates the selective nature of the antagonistic action and gives some quantitative estimates about the intensity of this activity. Further studies might require growing the antagonists in liquid culture and testing filtrates for antimicrobial activity. A final step could involve extraction of the antibiotic principle from the filtrate. In the following methods, the microorganism being tested for antibiotic activity is referred to as the "antagonist." The "test organism" is the microorganism at which this activity is directed.

A. AGAR CULTURE

Placing both a potential antagonist and test organism on the surface of a nutrient agar medium at a distance from each other is the most common method of determining antibiotic activity in pure culture. The organisms grow towards each other and one of the following reactions may be observed: (1) mutual intermingling of the two organisms; (2) inhibition of one organism on contact; the other organism continues to grow unchanged or at a reduced rate through the colony of the inhibited organism; (3) mutual inhibition on contact; the space between the two colonies is small, but clearly marked; (4) inhibition of one organism at a distance; the antagonist continues to grow through the resulting clear zone at an unchanged or reduced rate; or (5) mutual inhibition at a distance. Antagonism, with or without an antibiotic substance being produced, occurs when one or more of the following conditions take place:
1. Zone of inhibition develops between the two colonies. The antagonist may or may not continue to grow. If the growth of the antagonist is stopped, mutual antagonism is indicated. Thickness of the medium may influence the size of the clear zone.
2. After meeting, the hyphae of the fungal test organism die back and disintegrate, while the antagonist may continue to advance. If a

bacterial test organism is used, this type of antagonism is indicated by lysis of the bacterial colony.
3. Actual parasitism of fungal hyphae by the antagonist.
4. Flattening of the colony of the test organism on the sides nearest the antagonist.
5. Distinct general stunting or malformation of the test organism colony as compared with that on control plates in which the test organism is grown alone.

It should be pointed out that a clear zone of inhibition resulting between two colonies does not necessarily mean that an antibiotic substance has been produced. Hsu and Lockwood (1969) have evidence that some zones of inhibition result from nutrient deprivation of the test organism.

The rates of growth of the antagonist and test organism should govern the placement of the organisms on agar media. If both grow slowly, they can be placed close together (2-4 cm). If one grows faster than the other, they should be placed at the periphery of the dish opposite each other. If the test organism is a fast grower, such as many isolates of *Pythium* and *Rhizoctonia,* a slowly growing potential antagonist is usually placed at the periphery of the dish 2-5 days prior to application of the test organism. Following are some of the methods of placement that have been used with plant pathogens. Some examples are listed after each technique.

    1. The antagonist and test organism are placed on agar medium opposite each other at the periphery of a petri dish (Fig. 13-1).

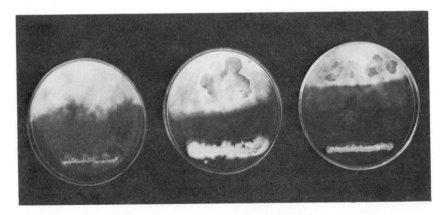

FIG. 13-1. Use of test method 1 (see text) for the detection of antagonism to *Pythium arrhenomanes.* Left to right: actinomycete, fungus, and bacterium. (Johnson, 1952)

e.g., bacteria, fungi, and actinomycetes vs. *Pythium arrhenomanes* (Johnson, 1954), *Rhizoctonia solani* (Boosalis, 1956), and *Sclerotium rolfsii* (Morton and Stroube, 1955).
2. The antagonist is placed on a medium in the form of a spot or streak. The test organism is placed 2-4 cm from the antagonist.
e.g., bacteria, fungi, and actinomycetes vs. *Fusarium oxysporum* f. *cubense* (Meridith, 1944), *Ophiobolus graminis* (Broadfoot, 1933), and *Verticillium dahliae* (Isaac, 1954).
3. The antagonist is placed on a medium at two places 4 cm apart. The test organism is placed midway between the two.
e.g., actinomycetes vs. *Alternaria solani* (Alexopoulos and Herrick, 1942).
4. The antagonist is placed on a medium at 3 equidistant places around the periphery of the petri dish. The test organism is placed in the center.
e.g., bacteria, fungi, and actinomycetes vs. *Botrytis cinerea* (Wood, 1951).
5. The antagonist is streaked on an agar medium at the periphery of the petri dish. The test organism is streaked at right angles to the original streak of the antagonist (Fig. 13-2).
e.g., bacteria fungi, and actinomycetes vs. *Agrobacterium tumefaciens* (Pridham, et al., 1956).
6. The antagonist is streaked or spotted on a medium at the periphery of the petri dish. Four days later an aqueous spore or mycelial suspension of the test organism is sprayed on the agar (Fig. 13-2).
e.g., bacteria, fungi, and actinomycetes vs. *Erwinia amylovora* (Stessel, et al., 1953) and *Fusarium oxysporum* f. *pisi* (Pridham, et al., 1956).
7. The antagonist is streaked or spotted on an agar medium. Five days later, 3 ml of an agar suspension containing the test organism are poured over the antagonist.
e.g., actinomycetes vs. *Streptomyces scabies* (Peterson, 1954) and *Fusarium culmorum* (Stevenson, 1956).
8. Agar medium containing spores and/or mycelial fragments of the test organism is poured into petri dishes. The antagonist is spotted or streaked on the surface of the seeded agar.
e.g., bacteria, fungi, and actinomycetes vs. *Erwinia carotovora* (Patrick, 1954) and *Helminthosporium sativum* (Anwar, 1949).
9. An agar suspension containing spores and/or mycelial fragments of the test organism is poured into petri dishes. A paper assay disc is placed on the surface of the hardened agar. One drop of a broth culture or a spore suspension of the antagonist is placed on each disc.

e.g., bacteria vs. *Pseudomonas syringae* (Teliz-Ortiz and Burkholder, 1960).

10. A section of unwaxed cellophane is placed in the bottom of a petri dish under an aluminum ring (1.6 cm high and 5.1 cm diam) into which 5 ml of an agar medium are added. The antagonist is placed on the agar surface in the form of a small mycelial transplant or a loop of cells or spores. The apparatus is inverted in the dish and an agar disc of the test organism is placed on the cellophane surface opposite the antagonist (Fig. 13-3).

e.g., bacteria, actinomycetes, and fungi vs. *Helminthosporium sativum* (Williams and Willis, 1962).

Two or more of the placement techniques described above may be equally effective for showing antagonism. A comparison of two test methods for determining the width of the zone of inhibition produced when an actinomycete was tested against *Verticillium albo-atrum* is shown in Fig. 13-2. The width of the zone was approximately the same in both cases. Probably the most important factor in the selection of a test method for a certain pathogen is the time lag between the application of the test organism and the antagonist. The selection of the proper time lag should depend primarily on the respective growth rates of the antagonist and test organism. Other factors that should be considered include type of agar medium used, depth of medium in petri dish, and temperature of incubation.

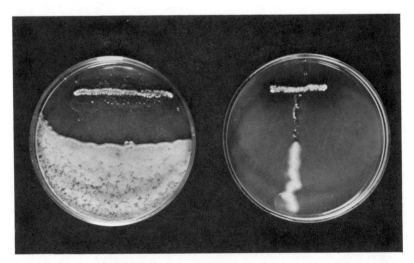

FIG. 13-2. A comparison of test methods 6 (left) and 5 (right). The antagonist is *Streptomyces* sp. and the test organism is *Verticillium albo-atrum*.

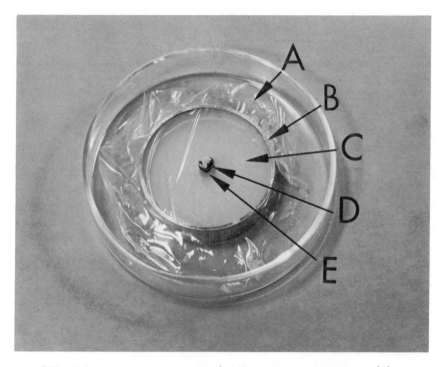

FIG. 13-3. Agar ring apparatus for determining antagonism. (A) sheet of unwaxed cellophane; (B) aluminum ring; (C) agar layer underneath the cellophane; (D) disc of assay fungus; (E) potential antagonist colony (on lower surface of agar.) (Williams and Willis, 1962. Phytopathology 52:368-369.)

B. ASSAY OF CULTURE FILTRATES OR ANTIBIOTICS

A liquid culture medium may be tested for microbial activity directly after removing the potential antibiotic-producing organism. The medium can be rendered sterile by heat, filtration, or centrifugation. To avoid possible inactivation of the active substance by heat, filtration through sintered glass, ultra-fine porcelain, or cellulose-membrane (Millipore) filters is commonly employed. Seitz filters are rarely used because they have been found to adsorb the active material from some antibiotic solutions. Selection of a suitable assay medium is important. Two main requisites are that the medium should support growth of the test organism and that it should not inactivate the antibiotic. Since constituents of the assay medium affect the degree of inhibition, standardization of the medium is a necessity for accurate and comparable determinations. For assaying commercial antibiotics a number of special media have been developed, such as Penassay Broth, Streptomycin Assay Agar, and Tyrothricin Assay Broth. Ingredients of a number of assay media that can be prepared in the laboratory are listed in Chapter 16.

Standardization of the cell concentration of the test organism is useful for precise assays. Cultures of bacteria or conidial suspensions can be diluted with water or physiological saline to a known concentration by using an electronic densitometer or photoelectric colorimeter. After obtaining a known concentration, the suspension may be further diluted to a standard previously determined as optimum for the assay.

Methods outlined below are designed for assaying sterile culture filtrates, extracts of culture filtrates, or commercial antibiotics.

1. **Broth-tube method (Waksman, 1945; Pratt and Dufrenoy, 1953).**

Decimal or serial dilutions of sterile culture filtrates or antibiotics are made in a series of broth tubes. Nutrients in the broth should be of such concentration as to compensate for dilution by the antibiotic solution. When sterile culture filtrates are tested for antibiotic activity, the protocol in Table 2 may serve as a guide. The protocol in Table 3 may be used when diluting solid antibiotics.

If bacteria are used as assay organisms, the inoculum should be light—one drop of a 1:10 or 1:100 dilution of a faintly turbid suspension or broth culture per tube. Controls should include antibiotic-free tubes inoculated with the assay organism, and others left uninoculated. The tubes are incubated at a proper temperature for 24 hr, or until growth occurs in the antibiotic-free inoculated controls. The lowest concentration of the test substance that completely inhibits the growth of the assay organism is regarded as the end point.

For precise determinations of the effect of antibiotics on growth of bacteria in liquid culture, turbidity measurements can be made following the method of Joslyn and Galbraith (1950). All dilutions are prepared in test tubes matched to read equally in a photoelectric colorimeter or densitometer. The assay organism is standardized by holding it in an ice bath after it has reached an optical density of 35% light transmission. It is diluted and standard amounts are added to each antibiotic dilution. When the light transmission of the culture control reaches 38%, all cultures are shaken and turbidity measurements are made. The percentage of growth in each dilution is calculated by dividing the optical density of the test dilution by the optical density of the control and multiplying by 100. The percentage of inhibition is determined by subtracting the percentage of growth from 100. Fifty percent inhibition is used as the end point.

Inhibition of mycelial growth of some fungi may be determined by the broth-tube method. Small discs of agar containing the mycelium of a fungus, such as *Pythium*, are placed in the bottoms of test tubes containing dilutions of the test solution. After incubation, vertical growth of mycelium is measured visually in ml (Johnson, 1953). Another method involves inoculating dilutions of

TABLE 2

A DILUTION-SERIES OUTLINE FOR DETERMINING THE ANTIBIOTIC ACTIVITY OF CULTURE FILTRATES ON MICROORGANISMS BY THE BROTH-TUBE METHOD

Dilution Series No.	Sterile water dilution	Parts of culture medium in parts of water	Volume of medium in tubes (ml)	Amount of dilution added to medium (ml)	Final conc.
1	None	1 in 0	0.0	5.0	1 in 0
2	"	1 in 0	2.5	2.5	1 in 1
3	"	1 in 0	3.3	1.7	1 in 3
4	"	1 in 0	4.0	1.0	1 in 5
5	3 ml in 3 ml*	1 in 1	4.0	1.0	1 in 10
6	3 ml in 3 ml**	1 in 2	4.0	1.0	1 in 20
7	"	1 in 4	4.0	1.0	1 in 40
8	"	1 in 8	4.0	1.0	1 in 80
9	"	1 in 16	4.0	1.0	1 in 160
10	"	1 in 32	4.0	1.0	1 in 320
11	"	1 in 64	4.0	1.0	1 in 640
12	"	1 in 128	4.0	1.0	1 in 1,280
13	Water only	0 in 1	4.0	1.0	Control

*3 ml of the full strength sterile filtrate added to 3 ml of sterile water.
**3 ml of the previous dilution added to 3 ml of sterile water.

TABLE 3
A DILUTION-SERIES OUTLINE FOR DETERMINING THE EFFECT OF SOLID ANTIBIOTIC SUBSTANCES ON MICROORGANISMS BY THE BROTH-TUBE METHOD

Dilution Series No.	Sterile water dilution	Grams of antibiotic in ml of water	Amount of dilution added to 4 ml medium (ml)	Final conc. (μg/ml)
1	1 g in 249 ml	1 in 250	1	800.0
2	5 ml in 5 ml*	1 in 500	1	400.0
3	,,	1 in 1,000	1	200.0
4	,,	1 in 2,000	1	100.0
5	,,	1 in 4,000	1	50.0
6	,,	1 in 8,000	1	25.0
7	,,	1 in 16,000	1	12.5
8	,,	1 in 32,000	1	6.3
9	,,	1 in 64,000	1	3.1
10	,,	1 in 128,000	1	1.6
11	,,	1 in 256,000	1	0.8
12	Sterile water only		1	0.0

*5 ml of the previous dilution added to 5 ml of sterile water.

test solution in Erlenmeyer flasks with the test fungus. After incubation (shaken or stationary), the contents of the flasks are filtered and the mycelium is washed and dried in an oven at 103 C. Net dry weight of mycelium produced is determined (Smith, 1960).

2. **Agar plate-dilution method (Waksman, 1945; Pratt and Dufrenoy, 1953).**

The antibiotic substance to be tested is serially diluted in sterile, distilled water. If the antibiotic substance is a solid material, the protocol in Table 4 may serve as a guide. The proper amount of the antibiotic solution is added to tubes, each containing 20 ml of sterile agar medium which has been melted and cooled to 42-45 C. The contents of each tube are shaken, poured into a sterile petri dish, and left to solidify. The antibiotic solution may be added to the melted agar medium after it has been poured into petri dishes, but it is difficult to obtain adequate dispersion.

A loop of a faintly turbid suspension of a bacterial assay organism is streaked on the surface of the agar. Several strains may be tested on each plate. The age of a bacterial test organism is very important—cultures 16-24 hr old usually are best suited for these tests. Time and temperature of incubation depend upon the type of organism used. For bacteria, 24 hr usually is sufficient

TABLE 4

A DILUTION-SERIES OUTLINE FOR DETERMINING THE EFFECT OF SOLID ANTIBIOTIC SUBSTANCES ON MICROORGANISMS BY THE AGAR PLATE-DILUTION METHOD

Dilution Series No.	Water solution of antibiotic ($\mu g/ml$)	Volume added to 20 ml agar (ml)	Final conc ($\mu g/ml$)
1	0	0.2	0.00
2	10	0.2	0.10
3	10	0.7	0.33
4	10	1.4	0.65
5	100	0.2	1.00
6	100	0.7	3.33
7	100	1.4	6.54
8	1,000	0.2	10.00
9	10,000	0.2	100.00
10	100,000	0.2	1,000.00

time. The highest dilution (the lowest concentration of the test substance) that inhibits growth of the test organism is regarded as the end point.

Generally, fungi are tested one to a plate (Gregory et al., 1952a). A disc (5-7 mm diam) of the test organism, cut with a cork borer from a culture, is placed on the surface of the medium containing the antibiotic. The dishes are incubated at 25-30 C and measurements are made after 24, 30, and 48 hr. The diam of the colony minus the diam of the original inoculum disc (5-7 mm) is a measure of linear growth. A smooth curve extrapolated to zero growth can be obtained by plotting the logarithm of the concentrations against linear growth. Concentration of antibiotic at this intercept is considered the end-point of the assay and equal to 1 unit.

3. The cylinder plate (Pratt and Dufrenoy, 1953; Waksman, 1945).

The cylinder-plate method is based on diffusion of the antibiotic materials into agar from steel, glass, or porcelain cylinders placed on agar media in petri dishes which have been seeded with the test organism. The medium is seeded by one of the following methods:

a. A 1:10 or 1:50 dilution of a 16- to 24-hr culture (if the test organism is a bacterium), or a conidial suspension of a fungus in sterile water, is flooded over the surface of the agar. Excess fluid is removed with a pipette.

b. One-tenth ml of a 1:10 dilution of a broth culture or a conidial suspension of the test organism is placed on the agar in the center of the dish. The inoculum is spread evenly over the surface of the agar with a sterile bent glass rod.

c. A base layer 2-3 ml deep is poured into the petri dish and left to solidify. One-tenth ml of a 1:10 dilution of the test bacterium or a conidial suspension is added to 5 ml of melted, cooled agar, and mixed thoroughly. This mixture is poured over the base layer and left to solidify.

Sterile, hollow cylinders, open at each end, and made from glass, porcelain, or steel are placed on the surface of the seeded agar and are filled with the test solutions. Commercially available stainless steel cylinders are 10 mm long with an inside diam of about 6 mm. The cylinders should be uniform in size and volume. Each dish of seeded agar may contain cylinders with different concentrations of the test solutions. After a period of incubation at a suitable time and temperature, zones of inhibition around the cylinders are measured to the nearest 0.5 mm.

Any suitable reservoir may be substituted for the cylinder provided that it permits diffusion of the antibiotic substance. For example, wells can be made by removing a plug of agar with a cork borer. These wells are then filled with the test solution.

4. The paper-disc method (Loo et al., 1945).

Twenty ml of sterile assay agar are added to petri dishes (Pyrex, 100 mm in diam, selected for uniform flat bottoms). After the assay agar has hardened, 4 ml of cooled agar seeded with the test organism are distributed evenly over the surface. The quantity of assay organism added to prepare the seeded agar should be determined by trial and adjusted so that a clear sharp zone of inhibition is obtained.

Filter paper discs (Schleicher and Schuell, 740 E, 1.3 cm) are placed flat side down on the seeded agar. A sample (0.08 ml) is pipetted immediately (within 5 sec) onto each disc. The tip of the pipette containing the sample is used to press the disc to the agar. Three to 6 discs may be placed symetrically around the center of the dish, not less than 10 mm from the edge. Each concentration of the antibiotic substance or culture filtrate should be assayed in 2 or more dishes. After incubation at the proper temperature and time suitable for the test organism, diameters of the zones are measured to the nearest ¼ mm (Fig. 13-4).

Paper discs previously impregnated with various concentrations of many antibiotics are available commercially.

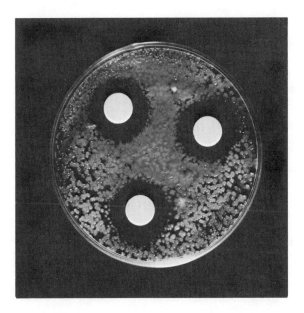

FIG. 13-4. Inhibition of *Pseudomonas tabaci* by Aureomycin with the paper-disc assay method. Each disc contained 30 µg of Aureomycin.

5. Gradient-plate technique (Szybalski, 1952).

This technique was originally designed to discover strains of microorganisms resistant to antibiotics, but it can be used to compare sensitivity of microbial strains to a specific antibiotic substance. Assay plates are prepared so that a gradual increase in antibiotic concentration is effected along one horizontal axis of each plate. This is accomplished by pouring two 20-ml layers of agar. The bottom layer consists of 20 ml of nutrient agar, left to harden with the plate slanted so that the entire bottom is barely covered. With the dish in the normal horizontal position, another 20-ml portion of agar containing the appropriate concentration of antibiotic is added (Fig. 13-5). Downward diffusion of the material, which becomes diluted in proportion to the ratio of thickness of the agar layers, establishes a uniform linear concentration gradient during subsequent incubation. The antibiotic may also be incorporated in the bottom layer, and by changing the ratio of its concentrations in both layers it is possible to adjust the range and the slope of the gradient, which remains surprisingly stable for several days after preparation of the plate.

Bacterial cells or conidial suspensions may be sprayed onto the surface of the agar or streaked along the horizontal axis of increasing concentration. Patterns of inhibition that may be encountered are illustrated in Fig. 13-6.

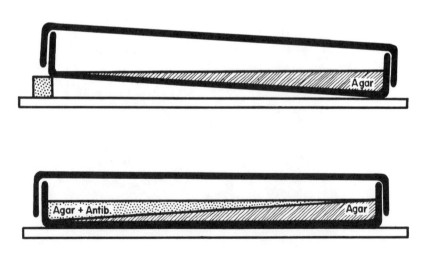

FIG. 13-5. Method of preparing a gradient plate. (Szybalski, 1952. Science 116:46-48.)

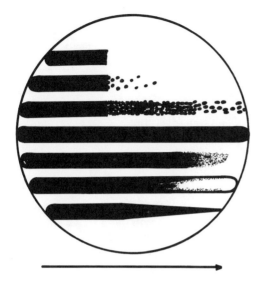

FIG. 13-6. Examples of inhibition patterns that may be found on gradient plates. (Szybalski, 1952. Science 116:46-48.)

A "double-gradient" agar plate was described by Weinberg (1957). The first layer may contain a compound such as a metalic salt and is solidified while the plate is in an inclined position. The second (upper) layer, which contains the antibiotic, is solidified while the plate is in a level position. Concentration of the antibiotic used should be determined with preliminary single gradient plates. In actual tests, single-gradient plates of both compounds serve as controls. By using several double-gradient plates over a range of concentration of both materials, the interaction between the compound tested and the activity of antibiotics can be measured.

6. Slide-germination techniques.

Methods familiar to plant pathologists for fungicide bioassays may also be used with antibiotics. A complete description of the standard slide-germination technique is found in Sharvelle (1961).

Gliotoxin can be assayed for activity to *Sclerotinia fructicola* with the following slide technique (Evans and Gottlieb, 1955). To 1 ml of various dilutions of the substance to be assayed is added 0.1 ml of a spore suspension of the fungus. Three separate drops of each dilution are placed on a slide and incubated in a moist chamber for 8 hr at a suitable temperature. Percent germination in each dilution is recorded. Number of units of antibiotic in the medium is expressed as the inverse of the dilution in which germination of 50% of the spores is prevented. All dilutions are prepared with a citric-sucrose solution (5.0 g sucrose and 0.005 g citric acid per liter); it permits good germination of spores of many fungi.

A unique assay slide-germination technique developed by Vasudeva et al. (1958) is used to test the effect of "bulbiformin" on *Alternaria tenuis*. Dilutions of the test liquid are prepared, and 0.02 ml of each is placed in each groove of a double grooved slide. The material is dried at room temperature. A spore suspension of the fungus is prepared (5000 spores/ml) and 0.02 ml of this suspension is placed over the dried drop of test liquid on the slide. After incubation for 18-24 hr in a moist chamber at 25-28 C, observations are made through the low power of a microscope. At low concentrations of the antibiotic the spores germinate, but characteristic bulbs form on germinating hyphae. The highest dilution at which bulb formation disappears is taken as the end-point for antibiotic activity. Inhibition and bulb formation are not due to two different substances, but are expressions of varying concentrations of the same substance.

CHAPTER 14
PRODUCTION, STABILITY, AND ACTIVITY OF ANTIBIOTICS IN SOIL

Soil is the source of most of the species of microorganisms that produce antibiotics. Yet antibiotic production in soil is difficult to demonstrate, except in sterilized soil and in soil supplemented with organic materials such as soybean meal, glucose, or oat straw. In the absence of an added nutrient source and in the presence of the natural soil microflora, antibiotic producing organisms either fail to grow, or antibiotics are produced in such small quantities that they are not easily detectable with present techniques of assay. Moreover, if antibiotics are formed, they may be immediately inactivated in natural soil. Purified antibiotics incorporated into soil may be adsorbed on soil colloids, decomposed by the actions of the soil microflora, inactivated by chemicals present in soil, or they may be unstable at the acid or alkaline reaction of the soil. Despite the ultimate fate of antibiotics in soil, some are more persistent than others. For example, oleandomycin and erythromycin retain activity for 60-70 days, whereas penicillin is active for only 2 days after application (Kruger, 1961).

Most antibiotics, the fate of which has been studied in soil, are commercial chemicals developed for control of human and animal diseases. It is possible that the stability in soil of yet unknown antibiotics active against plant disease fungi and bacteria might be different from these commercial ones. The question has arisen as to whether antibiotic production in natural soil affects the survival of susceptible microorganisms, including plant pathogens. Menzies (1963) concluded that antibiosis is a factor in destroying pathogens in soil, but insufficient evidence is available to judge its importance. With special techniques, production of only a few antibiotics in nonsterile, unamended soil has been demonstrated (Hessayon, 1953; Rangaswami and Ethiraj, 1962; Mirchink and Greshnykh, 1962). From such demonstrations and from the voluminous accumulated data, which indirectly support the thesis of antibiotic production in soil, it is reasonable to believe that antibiosis is an important factor affecting the activity and survival of microorganisms in natural soils. Brian (1957) envisions that antibiotics are produced principally in microenvironments where suitable substrates occur. Such microenvironments might be fragments of organic debris, seed surfaces, or rhizospheres. It would appear that antibiotics are not produced continually in these microenvironments, but only when temperature, moisture, and availability of nutrients are at optimum levels.

A. RECOVERY OF ANTIBIOTICS ADDED TO SOIL

Persistence or inactivation of antibiotics added to soil can be determined by extracting the antibiotic from soil with solvents, concentrating the extracts, and testing the partially purified material for biological activity. The particular extraction procedure depends on the nature of the antibiotic used. The following three techniques are quite efficient for extracting certain antibiotics from soil and can be used as guides. Bioassay techniques for antibiotics extracted from soil are found in Chapter 13.

1. Chloromycetin (Gottlieb and Siminoff, 1952).

Twenty-g samples of antibiotic-containing soil, adjusted to 60% of its moisture-holding capacity, are extracted 3 times on a reciprocal shaker with 25, 15, and 10 ml of methanol for 25, 15, and 10 min respectively. The extracts are filtered under suction through two layers of Whatman No. 1 filter paper in a Buchner funnel and the residues washed 3 times with methanol. The combined filtrates are evaporated to dryness in a forced-draft oven at 70 C. Finally, the dry residue is taken up in 5 ml of 10% methanol and assayed for chloromycetin. This elution technique recovers 70-80% of the antibiotic contained in the soil.

2. Gliotoxin (Evans and Gottlieb, 1955).

Ten-g samples of soil containing the antibiotic are extracted with 10-ml quantities of ethyl acetate, followed by three consecutive extractions with chloroform. All extracts, including the ethyl acetate, are transferred to a flask and evaporated to dryness under reduced pressure at room temperature. The solid residue is dissolved in 5 ml of 10% ethyl alcohol, and dilutions of this solution are used for assay. This technique recovers 52-90% of the antibiotic added to soil.

3. Extraction with buffers (Soulides, 1964).

Treating relatively large amounts of soil with suitable buffers releases certain antibiotics adsorbed on soil particles. Minute quantities of soil-adsorbed antibiotics can be detected with proper use of buffers. Two ml of buffer solution per gram of soil are added to 500-g samples of soil containing the antibiotic. The samples are shaken at least 2 hr on a mechanical shaker. The buffer solutions consist of mixtures of K_2HPO_4 and KH_2PO_4 adjusted to pH 8.0 for streptomycin and Carbomycin, pH 5.0 for Terramycin and Aureomycin, and pH 6.0 for Bacitracin. The extract is separated by centrifugation followed by filtration. The soil sediment is resuspended in the same buffer solution, then centrifuged a second time, and the supernatant is filtered. This is repeated twice and the sediment is transferred into a Buchner funnel provided with Whatman filter paper No. 42 and washed with buffer solution. The collected extract amounting to about 3 ml/g of soil is flash evaporated at a low temperature (25-40 C). The resulting dry residue is repeatedly treated with a solvent in which the antibiotic is highly soluble and the salt residue practically insoluble. Solvents used were methanol for streptomycin, Aureomycin and Bacitracin, and acetone for Carbomycin and Terramycin. The solvent is expelled by evaporation *in vacuo*. The resulting residue is dissolved in 1.0 ml of the original buffer, brought up to pH 8.0 for streptomycin and Carbomycin, and to 6.5-6.8 for the remaining antibiotics. To determine the efficiency of extraction, these solutions are bioassayed and compared with controls.

B. ISOLATION OF ANTIBIOTICS PRODUCED IN SOIL

Some microorganisms produce antibiotics in soil amended with organic matter. Antibiotics produced are extracted from soil, bioassayed for antibiotic activity, and compared to the antibiotic produced in pure culture. For methods of bioassay in pure culture, see Chapter 13.

1. Extraction with organic solvents.

   a. **Gliotoxin (Wright, 1952).** Nonsterilized soil supplemented with organic matter is inoculated with a spore suspension of a gliotoxin-producing strain of *Trichoderma viride* and incubated at 25 C for 4 days. The soil is extracted with ether and the extract used to demonstrate and identify the antibiotic. Identification can be demonstrated by a bioassay based on paper chromatography. Chromatograms are run with the soil extracts and with solutions of pure gliotoxin on filter paper strips. These are placed on agar seeded with *Bacillus subtilis*. After incubation, the distance moved by the active substance is demonstrated by the position of a clear zone of inhibition. R_f values of pure gliotoxin and soil extracts are compared. The amount of gliotoxin present can be estimated roughly from the size of the inhibition zone.

   b. **Bulbiformin (Vasudeva et al., 1962).** Sterilized or nonsterilized soil supplemented with organic materials (dextrose, compost, clover) is inoculated with a strain of *Bacillus subtilis* which produces bulbiformin. After incubation, 25-g quantities of soil are extracted with 20 ml of water-saturated n-butanol.

After intermittent agitation for 2½ hr, the solvent is recovered by centrifugation at 2,600 rpm for 10 min. Dilutions of the clear solvent are assayed with *Alternaria tenuis* as the test organism. At high concentrations, growth of *A. tenuis* is completely inhibited, but characteristic bulb formation occurs on young hyphae of germinating spores exposed to lower concentrations of the antibiotic. See Chapter 13 for details of the slide test used.

2. Use of soil solutions.

    a. **Centrifugation (Grossbard, 1952).** Previously sterilized soil is inoculated with *Penicillium patulum*. After incubation the soil is subjected to high-speed centrifugation and the fluids obtained are assayed for patulin with sensitive bacteria. Maximum antibiotic production can be obtained when glucose and organic residues have been incorporated into the soil.

    b. **Displacement (Gregory et al., 1952a).** Sterilized soil supplemented with organic materials is inoculated with the antibiotic-producing microorganism. After incubation, the soil solution is displaced with 95% ethyl alcohol (see Chapter 8 for procedure). The soil solution obtained is assayed by the paper-disc method for antibiotic activity.

    c. **Percolation of soil (Stevenson and Lochhead, 1953).** Ths technique makes use of the soil-perfusion apparatus described by Audus (1946) (Chapter 8, Fig. 8-1). Briefly, the method consists of automatic and intermittent perfusion of a solution through a column of soil. The activity of the selected antagonists inoculated into the soil column is calculated from periodic assays of the perfusing solution for the presence of antibiotic substances. The soil column contains 50 g of the 2- to 3-mm sieve-fraction of air-dried soil. Solid organic material may be added and mixed with the soil if desired, and the soil can be perfused with distilled water. In the event that a water-soluble carbohydrate is to be added, the soil is percolated with 100 ml of a solution of this material held in the reservoir.

For experiments requiring sterile soil, the soil column is autoclaved for 45 min at 15 lb pressure. Then the soil column is attached to the remainder of the apparatus and the whole apparatus is further autoclaved for 15 min. The percolate is sterilized separately, and 100 ml are added aseptically through the sampling tube to the reservoir F. Sterility of the apparatus is insured throughout the experimental period by a cotton filter fitted to the sample tube. Inoculation of the soil is accomplished by loosening the rubber stopper at the top of the soil column and introducing aseptically 1 ml of a suspension of the microorganism to be studied. Operation of the unit is started prior to inoculation, and the percolate is perfused through the soil until it is completely wet. This insures even distribution of the inoculum throughout the soil column. The percolate is assayed daily for antibiotic activity.

3. Extraction with chromatograms (Mirchink and Greshnykh, 1962).

Soil was inoculated with *Penicillium cyclopium* or *P. purpurogenum*. Antibiotics were produced in both sterilized and nonsterilized soil, either amended or not amended with organic matter. The following technique was used to compare antibiotic substances formed by the organism in soil with those formed by the same organism in pure agar culture.

Small clumps of moist soil containing the antibiotic substance are placed on the ends of paper strips. They are removed and additional clumps are placed repeatedly on the strips. This process is also performed with small blocks of agar containing the antibiotic-producing organism. The paper strips are dried at room temperature and the ends immersed in suitable solvent systems, e.g., acetone, benzol, sulfuric ether, butyl-acetate, chloroform, butanol, benzol-acetic acid-water (2:2:1), or butanol-acetic acid-water (2:1:1). When the solvent has traveled a distance of 25 cm, the strips are removed, dried, and placed on the surface of agar inoculated with the test organism. The position of the substance on the chromatographic paper is determined after incubation by the position of zones of inhibition on the assay plates. If the same inhibition pattern occurs on plates containing strips treated with soil and on plates containing strips treated with agar, it is probable that the antibiotic produced in both substrates is identical.

C. DIRECT BIOASSAY OF SOIL FOR PRESENCE OF ANTIBIOTICS

Although extraction before bioassay is valuable in determining the stability of antibiotics, it provides no information about biological activity of the antibiotic in the soil under conditions of the experiment. Direct bioassays yield information on stability, persistence, and inactivation of antibiotics in soils. The following techniques have been used with success for measuring these factors. In all cases, soils without added antibiotics or antibiotic-producing microorganisms should be used as controls. When measurements of zones of inhibition serve as criteria for the activity of the antibiotic in soil, a series of known concentrations of the antibiotic in the absence of soil should be tested in parallel. A graph showing the relation between the size of the zone of inhibition and the concentration of the antibiotic is made and should be referred to when determining concentrations of the antibiotic in soil.

1. Population assay (Gottlieb and Siminoff, 1952).

The antibiotic (or antibiotic-producing microorganism) is added to sterilized soil previously inoculated with the test organism. After incubation, population of the test organism is determined, usually with the dilution-plate method. Magnitude of this population is compared to that of the control soil which has received no antibiotic.

2. Use of agar culture.

   a. **Cylinder plate (Witkamp and Starkey, 1956).** Steel or glass cylinders measuring 6-7 mm in diam, and containing saturated soil previously sterilized and treated with the antibiotic, are placed directly on assay agar seeded with the test organism. The dishes are refrigerated at 2 C for 5 hr for diffusion of the antibiotic into the test substrate. After incubation at the proper temperature, zones of inhibition are measured.

   b. **Soil-agar slurry (Witkamp and Starkey, 1956).** A soil-agar slurry is made by dispersing 5 g of antibiotic-treated sterilized soil in 8 ml of the assay agar medium. When the slurry has solidified, it is streaked with the test organism. Growth of the test organism is compared to that on plates without the soil.

   c. **Layering (Witkamp and Starkey, 1956).** Five g of sterilized, antibiotic-treated soil are placed in petri dishes and saturated with 4 ml of water agar. The material in each dish is covered with 4 ml of nutrient agar and inoculated with the test organism. Growth of the test organism is compared to that on similar layers without soil.

   d. **Hole plate (Kruger, 1961).** A suspension of the test organism is mixed with cooled liquid agar medium. Ten ml of this mixture is poured into each petri dish in which a base layer of 10 ml of agar has been applied previously. The dishes are incubated overnight at 3-4 C. Holes of 10-mm diam are then cut with a cork borer connected to a suction pump. Holes thus made are filled carefully with antibiotic-treated previously sterilized soil. After incubation, zones of inhibition are measured.

3. Use of agar culture with barriers.

   Semi-permeable barriers, placed between soil and a test organism growing on an assay medium, aid in determining persistence of antibiotics and antibiotic production in nonsterile soil. Since some barriers are more permeable to antibiotics than others, and since different antibiotics vary in their rate of diffusion through the same type of barrier, it is advisable to run preliminary trials with different barriers and procedures.

   a. **Soil strip (Witkamp and Starkey, 1956).** Five g of soil containing the antibiotic are placed in a mold $8 \times 1 \times 0.5$ cm deep, and saturated with 2% agar. The soil strip thus formed is placed carefully on a cellulose film which covers part of an assay agar plate. Such assay plates are held at 2 C for 5 hr to permit diffusion of the antibiotic into the test substrate. Each film with attached soil then is removed, and the agar is inoculated with the test organism. The zone of inhibition is measured from the edge of the place previously occupied by the strip.

b. **Cylinder plate (Mitchell, 1963).** Petri dishes containing agar medium are spray-inoculated with the test organism. A sterile membrane filter (Millipore Corporation, Bedford, Mass.), with 0.22-μ pore size, is placed on the agar surface, and an aluminum assay cup is attached to the filter with paraffin. A 0.5-g sample of soil containing the antibiotic or the antibiotic-producing organism is placed in the assay cup. After incubation, zones of inhibition are measured.

c. **Fritted-filter tube (Koike and Gainey, 1958).** The tubes utilized are glass pyrex tubes, 200-mm overall length and 35 mm inside diam, in the center of which have been sealed fritted filters of ultrafine porosity with 0.9-1.4 μ maximum diam of pores (Fig. 14-1). The tubes are plugged with nonabsorbent cotton and steam sterilized prior to use. Approximately 2 ml of a suitable agar medium (a 2-3 mm layer) are placed on one side of the filter while the tube is held in a vertical position. After the medium has hardened the tube is inverted and 20 g of saturated soil containing the antibiotic are placed on the opposite side of the filter. Filters thus prepared are placed in a refrigerator at 2-5 C for 18-24 hr. Then the agar medium is inoculated with the test organism and after incubation, growth is compared with that in tubes containing soil with different concentrations of the antibiotic.

D. SPECIAL TECHNIQUES FOR DEMONSTRATING ANTIBIOTIC PRODUCTION IN SOIL

1. Soil-sandwich technique (Hessayon, 1953).

Trichothecin production in soil can be demonstrated with this procedure. Since the method requires that the material be autoclaved, the production of heat-stable antibiotics only may be studied. It was found that trichothecin was produced in both sterilized and nonsterilized soil.

Sixty-g layers of soil are placed in petri dishes and inoculated with a washed culture of the antagonist. Controls may consist of dishes of soil inoculated in a similar manner with spores and hyphae of the antagonist previously killed by autoclaving. Thus, allowance is made for minute amounts of the antibiotic in the inoculum. The soil plates are maintained at a specified moisture content and temperature throughout the course of the experiment. After incubation, the soil in each dish is mixed thoroughly and divided into four portions. A shallow layer of each portion is placed in a separate petri dish and sterilized by autoclaving. Ten ml of an appropriate agar medium are added to each dish to anchor the soil layer to the base of the dish. After solidification, another 20 ml of melted and

Control   50 ppm Actidione   100 ppm Actidione
8-17-54

FIG. 14-1. The fritted-filter tube method of testing antibiotics in soil. Note slight retardation in growth of *Pythium graminicola* Subr. on agar surface opposite soil containing 50 ppm cycloheximide (Acti-dione) and the complete inhibition in the tube with 100 ppm of the antibiotic. (Koike and Gainey, 1958. Soil Science 86:98-102. © 1958, The Williams & Wilkins Co., Baltimore.)

cooled agar medium are added to form a smooth surface layer of nutrient agar. The dishes are left to stand for 20-24 hr to permit diffusion of any antibiotic from the soil into the agar.

The agar may be assayed for antibiotic content by inoculation with an appropriate test organism. Lysis of mycelium or bacterial spores, or a decrease in the rate of growth of colony diameter may indicate the presence of an antibiotic. Since trichothecin in small concentrations stimulates growth of *Fusarium oxysporum* f. *cubense,* an increase in colony diam over that in control dishes indicates trichothecin production.

2. **The buried slide (Stevenson, 1956a).**

The buried-slide technique is essentially the same as that used by Chinn (Chapter 10) for studying the effects of soil conditions on germination of fungal spores. In this instance, modifications of the technique are used to study the production of antibiotics by microorganisms in soils through their effects on germination and on morphology of the developing hyphae of the test organism. Slides are prepared with spores of the test fungus in a thin agar film and inserted in pots containing soil previously sterilized and inoculated with antagonistic microorganisms. Slides removed at intervals of 1, 3, and 7 days are stained and examined for spore germination.

A modification of the above procedure consists of preincubating slides containing spores of the test fungus for 24 hr in a moist chamber at 26 C before they are placed in the antagonist-inoculated soils. This permits germ tubes to develop prior to insertion in soil. Slides removed from the soil at intervals of 1, 3, and 7 days are stained and examined for typical antibiotic effects such as growth inhibition, distortion, formation of protuberances, excessive branching, and lysis.

CHAPTER 15
BIOLOGICAL CONTROL

A broad concept of biological control of plant disease includes any type of disease reduction or decrease in inoculum potential of a pathogen brought about directly or indirectly by other biological agencies. Through common usage of the concept, man is exempted as a direct contributor. He can, however, manipulate conditions which favor activities leading to biological control.

Under the broad concept, addition to soil of naturally produced chemicals toxic to pathogens would constitute biological control. Thus, amending soil with alfalfa meal which contains saponins that inhibit growth of *Phytophthora cinnamomi* (Zentmyer and Thompson, 1967), and growing asparagus which contains a substance in its roots and root exudates toxic to *Trichodorus christiei* (Rohde and Jenkins, 1958), are examples of biological control. Most of the basic research on biological control of soil-borne pathogens, however, has been centered around attempts to utilize competitive activities of the saprophytic soil microflora. Although the nature and extent of microbial competition are not clearly defined, it is generally accepted that one organism affects activities of another when they are competing for space, oxygen, water, and nutrients. A fast-growing organism, with the capacity to produce antibiotic substances, and which is tolerant to antibiotics produced by other organisms, could be a strong substrate competitor. Some organisms are favored by their capacity for predation or direct parasitism on other organisms. Metabolites of some microorganisms may be utilized by others, resulting in stimulation and enhanced competitive nature of the latter. Nutrients released from dead cells may be selectively utilized as food materials. The extent and intensity of these activities are further governed by physical and chemical factors in the soil environment. To be successful, a pathogen must possess the necessary physiological characteristics for competition during the saprophytic phase of its life cycle. Man, however, may manipulate the environment through physical, chemical, or microbial modifications, so that stress is placed on the pathogen, with a resulting decrease of its competitive saprophytic ability and corresponding decrease in inoculum potential.

Research in recent years on the ecological and physiological activities of organisms in soil has made biological control an integral part of the broad field of plant pathology. However, because of the extremely complex nature of the soil microcosm, little of this work has led to practical application by growers. Several lines of approach have been made and the following is a discussion of some of the methods under study.

A. USE OF PURE CULTURES OF ANTAGONISTIC ORGANISMS

1. Soil infestation with antagonists.

Severity of some diseases can be reduced when pure cultures of antagonistic organisms are applied to previously sterilized, artificially infested soil (Fig. 15-1). Steps in the procedure usually consist of (1) infesting sterilized soil with a pathogen, (2) inoculating the infested soil with the test organism (antagonist), (3) growing susceptible plants in the soil, and (4) determining disease severity as compared with plants in soil containing only the pathogen and in soil containing only the antagonist. A time lag of a few days between each of these steps may or may not be necessary. Sources of inoculum of the antagonist may be aqueous spore suspensions, washed mycelial suspensions from liquid cultures, blended mixtures of mycelia, spores, and broth of liquid cultures, chopped agar cultures, or ground-up cultures of the organisms grown on plant materials such as wheat grains. Effects of added nutrients must be considered when evaluating results. The antagonist should be mixed thoroughly with soil containing the pathogen.

The test organism does not necessarily have to be antagonistic in agar culture for control to be obtained in soil (Broadfoot, 1933). Of 21 cultures of bacteria and fungi that controlled *Ophiobolus graminis* (take-all disease of wheat) in soil, only 15 were antagonistic in culture; of 45 that gave intermediate or no control in the soil, 17 were compatible and 28 were antagonistic. Other diseases can apparently be controlled only with antagonists. Only organisms antagonistic to the pathogen in agar culture were found by Johnson (1954) to control *Pythium* root rot of corn. Also, there was evidence of a positive relationship between the decrease in severity of disease in soil and the magnitude of the zone inhibition of the casual fungus in petri dish culture (Johnson, 1954; Stevenson, 1956).

Predators and parasites may be effective control agents. Trapping and killing of plant parasitic nematodes in soil by a number of fungal species have been observed, but the addition of predaceous fungi to root-knot infested soil has not been a successful control measure (Mankau, 1961; Hams and Wilkin, 1961). Rhoades and Lindford (1959) reduced infection of corn roots by *Pythium arrhenomanes* when larvae of the mycophagous nematode, *Aphelenchus avenae*, were added to *Pythium*-infested soil.

FIG. 15-1. Control of corn root rot caused by *Pythium arrhenomanes* in previously sterilized-infested soil. Left to right: corn from pathogen-infested soil, infested soil plus antagonistic actinomycete, and noninfested soil. (Johnson, 1954. Phytopathology 44:69-73.)

A few successful attempts to control diseases in nonsterile, naturally infested soil have been reported. Slagg and Fellows (1947) reduced infection by *Ophiobolus graminis* when *Aspergillus flavus* was added to soil naturally infested with the pathogen. Tveit and Moore (1954) found that antagonistic isolates of *Chaetomium* added to naturally or artificially infested nonsterile soil reduced infection of oat seedlings caused by *Helminthosporium victoriae*. These two reports were the results of laboratory and greenhouse tests where large quantities of the antagonists were added to relatively small quantities of soil; addition of such quantities of antagonists in the field would be impractical. When pure cultures of antagonists are added to soil along with large quantities of their organic substrate such as oat grains or cornmeal, considerable reduction in population levels of the pathogen are often obtained. The organic substrate itself may be partly responsible for the control effect; this will be discussed further in Section B of this chapter.

Promoting growth of antagonistic organisms in nonsterile soil is basic to practical biological disease control. To "establish" the antagonist at population levels high enough to produce the desired effect, a change in population of the existing microflora must occur, or the organism being added must have additional selective nutrients to overcome the fungistic or bacteriostatic effects of the natural microflora. Antagonists produce inhibitory by-products only under certain environmental conditions. The nature and constituents of the substrate on which they are grown have a direct bearing on the quantity and quality of metabolites produced. Some fungi produce substances inhibitory to *O. graminis* on one culture medium and stimulatory substances on another. Even on the same substrate, inhibitory substances may be produced by one fungus and stimulatory substances by another (Slagg and Fellows, 1947). Thus, establishing and/or promoting growth of an antagonist in soil may not necessarily favor reduction of the susceptible pathogen, but may actually increase severity of the disease.

2. **Treatment of seed with microorganisms.**

Control of some pathogens can be accomplished by treating seed with cultures of certain organisms. The general procedure is to apply bacterial cells, fungal spores, or mycelial fragments to seed coats, plant the seed in pathogen-infested soil, and later assess disease severity and compare it with controls of untreated seed. The usual method for applying the organism to the seed is to soak the seed in aqueous bacterial, spore, or mycelial suspensions. Some investigators dip the seed momentarily in the suspension; others soak them up to 36 hours. Usually, the seed are left to dry before sowing. Other methods include shaking seed in bottles in which sporulating cultures are growing on agar media (Wright, 1956), and shaking the seed in a 3% solution of sodium carboxymethyl cellulose containing spores, mycelial fragments, or bacterial cells (Tveit and Wood, 1955). Some reports of successful control are listed in Table 5.

3. Use of culture filtrates or antibiotics.

Addition of antibiotics or culture filtrates to soil for control of root diseases has met with little success. Many antibiotics incorporated into soil are adsorbed onto soil colloids, decomposed by the natural soil microflora, or inactivated by chemicals present in soil. Some are unstable at the acid or alkaline reaction of the soil. Others may be more stable or persistent in soil, but the large quantities needed to produce beneficial effects make their use impractical.

Some positive results have been obtained when plants or plant parts were soaked in antibiotic solutions before being placed in infested soil. Some reports of control are given in Table 6.

B. ORGANIC AMENDMENTS

Altering the nutrient status of soil by adding decomposable organic amendments may cause a major change in the microbial population. Some

TABLE 5

REPORTS OF CONTROL OF PLANT DISEASES BY TREATING SEED WITH MICROORGANISMS

Host	Pathogen	Seed treatment organism	Reference
Flax	*Fusarium, Colletotrichum*	Several soil bacteria	Novogrudsky et al., 1937
Wheat	*Helminthosporium sativum*	Several seed bacteria	Ledingham et al., 1949
Oat	*Fusarium nivale*	*Chaetomium* spp.	Tveit and Wood, 1955
Mustard	*Pythium* sp.	*Trichoderma viride*	Wright, 1956
Corn	*Fusarium roseum*	Bacterium, *Chaetomium* sp.	Kommedahl and Chang, 1966
Beet	*Pythium ultimum*	*Trichoderma* sp., *Penicillium* sp.	Liu and Vaughan, 1965
Tomato	*Pythium debaryanum*	*Arthrobacter* sp.	Mitchell and Hurwitz, 1965

components of the microflora are stimulated and others suppressed; a desirable effect is obtained if plant pathogens are suppressed. Many reports have been made of the effect of amendments on plant disease severity. One of the first

TABLE 6
REPORTS OF USE OF ANTIBIOTICS OR CULTURE FILTRATES OF ANTAGONISTIC MICROORGANISMS FOR CONTROL OF SOIL-BORNE PLANT PATHOGENS

Host	Pathogen	Method	Reference
Tomato	*Fusarium oxysporum* f. *lycopersici*	Soaked whole plant in solution of thiolutin, then transplanted into infested soil	Gopalkrishnan and Jump, 1952
Tomato, et al.	*Agrobacterium tumefaciens*	Soaked galls in solution of penicillin	Brown and Boyle, 1945
Tomato	*Corynebacterium michiganense*	Soaked roots 6 hours in solution of albamycin, then transplanted into infested soil	Kruger, 1959
Barley	*Xanthomonas translucens*	Soaked infested seed in solution of subtilin	Goodman and Henry, 1947
Sugar beet	*Aphanomyces cochlioides*	Soaked seed in solution of streptomycin, then planted in infested soil	McKeen, 1949
Oat	*Fusarium nivale*	Soaked seed in culture filtrate of *Chaetomium* sp., then planted in infested soil	Tveit and Wood, 1955
Peanut	*Aspergillus niger, Rhizopus*	Soaked infested seed in solutions of filipin or thiolutin	Kruger, 1960

reports was by King et al. (1934) who found that heavy manuring of cotton land either prevented the development of *Phymatotrichum* root rot, or delayed it until a cotton crop was matured. They suggested that a condition unfavorable to development of disease was produced in the soil by the presence of high populations of soil microorganisms.

Organic amendments in the form of crop plant residues have been used successfully to control root disease; generally better results have been obtained with mature-dry residues than with green residues. Many organic compounds, other than common crop residues also have been tested. These include glucose and dextrose, chitin, soybean and cottonseed meal, various kinds of animal manures, cellulose powder, sawdust, peat moss, powdered or chopped agar, ammonia, urea, and other pure organic compounds. Usually, amendments added to soil are more effective if they are chopped into particles 3 mm or smaller in size. For example, finely ground residues resulted in better control of pinto bean root rot than did coarsely chopped residues (Maier, 1959). The amendment should be mixed intimately with the soil. In greenhouse experiments this can be done by mixing the amendment and soil in a portable cement mixer or similar mechanical device. For field experiments the amendments are often rototilled thoroughly into the top 15-20 cm of soil. The incubation period (time between incorporation of the amendment and planting of the test crop) may be critical and varies with the type of amendment used and the pathogen. Addition of chitin to soil reduced *Fusarium solani* f. *phaseoli* infection of bean roots when seed were planted 0, 1, or 2 weeks after amendment incorporation (Mitchell and Alexander, 1962). No significant reduction in disease severity occurred with longer periods of incubation. Conversely, the effectiveness of amendments in reducing severity of *Rhizoctonia* disease of bean was greatest when beans were planted 3-7 weeks after amendment incorporation (Papavizas and Davey, 1960). The optimum incubation period for control of other diseases might be even longer and should be determined by trial.

To reduce root diseases, addition of relatively high rates of crop residues is usually required; rates of 1-2% by weight have yielded the most consistent results. This is roughly equivalent to 10-20 tons/acre-furrow-slice and probably would be impractical for large scale field operations. Smaller quantities may be required of organic amendments other than crop residues. Chitin, at the rate of 200-500 lb/acre, significantly reduced the severity of *Fusarium* root rot of beans (Mitchell and Alexander, 1961); urea, added to soil at the rate of 0.2%, completely inhibited germination of sclerotia of *Sclerotium rolsfii* (Henis and Chet, 1968).

Other factors which have been found to influence the effectiveness of disease control by the addition of organic amendments include (1) nature of the amendment, (2) C:N ratio of the amendment, and (3) soil environmental conditions such as temperature, moisture, and pH. These factors and mechanisms involved are discussed in detail in a recent summary (Cook, 1969).

C. CROP ROTATION

Crop rotation is a method of biological control if changes that occur in the soil microbial population affect a pathogen and reduce disease severity. This is a logical assumption when it can be shown that certain nonsusceptible crops are more effective than others in a rotation sequence (Curl, 1963). Hildebrand and West (1941) presented evidence that the incidence and severity of strawberry root rot could be modified by turning under certain cover crops, and that these modifications were correlated with changes in the microbial population of naturally infested soil. Severity of root rot was less after soybean rotation than after red clover or timothy. Deems and Young (1956) found that black root of sugar beets caused by *Aphanomyces cochlioides* was much less severe following corn and oats than after alfalfa or sugar beets. The total population of fungi, as determined by the dilution-plate method, varied little with the crops, but significantly more isolates of *Aspergillus fumigatus, Trichoderma viride*, and species of *Penicillium* were obtained from corn and oat soils.

Evidence that soybeans, used as a cover crop in rotation with potatoes, was associated with reduced incidence of scab was presented by Lochhead and Landerkin (1949) and Rouatt and Atkinson (1950). Actinomycetes were isolated from the rhizosphere of potatoes, and isolates selected at random were tested for antagonism to *Streptomyces scabies*. Many more antagonistic isolates were obtained from the soybean soils than from the control soils. A more recent study by Weinhold and Bowman (1965) revealed that a soybean cover crop and green manure incorporation prevented the buildup of scab, whereas barley, employed in the same manner, increased disease incidence. A predominant bacterium found in the plots produced an antibiotic similar to bacitracin. The bacterium when grown on a water extract substrate of soybean produced three times more antibiotic than when grown on a comparable extract of barley. Similar results were obtained when extracts of partially decomposed tissue were used.

Turning under cover crops or other crops in rotation produces effects similar to those of physically adding green manure amendments to soil. It has been shown that in most cases the amount of amendment required to reduce disease severity is larger than the amount required to reduce disease severity by the same crop in rotation. It would appear that release of root exudates into soil, the buildup of certain rhizosphere organisms, or the increased production of antibiotics could be partly responsible for the beneficial effects of some of the rotations cited above. A study of fungal populations in soils rotated with various crops (Williams and Schmitthenner, 1962) revealed that rotation resulted in a richer and more variable soil microflora than in soil monocropped with these same crops. Sixty-six groups of fungi were significantly affected in the rotation as contrasted to 30 in the monocropped plots.

D. STERILIZATION AND "PASTEURIZATION" OF SOIL, AND STIMULATION OF ANTAGONISTS BY CHEMICALS OR CULTURAL PRACTICES

Sterilization of soil can be a method for biological control within the framework of the concept that, after sterilization, the dominant microbial recolonizers are antagonistic to potential pathogens. The first recolonizers can increase in previously sterilized soil to a very high level because of the release of nutrients and reduced competition. A pathogen introduced into such soil by air contamination (or by other means) would find conditions unsuitable for growth and survival. A classic study to support this concept was made by Ludwig and Henry (1943) who reinfested sterile soil with *Ophiobolus graminis* and recontaminated it by additions of small amounts of field soil. *Trichoderma viride*, antagonistic to *O. graminis*, quickly became the dominant organism and the total microflora increased in density to a higher level than in nonsterilized soil. These increases in microbial activity resulted in less severe infection of wheat by *O. graminis* in recontaminated soil than in nonsterilized soil. Another approach to elucidation of the concept was made by Johnson (1952) in studies of root rot of corn caused by *Pythium arrhenomanes*. Very little root rot resulted when nonsterilized soil was inoculated with *Pythium*, but severe root rot occurred when sterilized soil was used. In order to determine which segment of the natural soil microflora effectively prevented disease development, soil was sterilized, infested with *Pythium*, and allowed to become recontaminated by exposure to the atmosphere. At 5-day intervals after sterilization, disease severity was assayed and microbial analyses were made to determine antagonistic reactions to *Pythium*. The severity of root rot decreased progressively as the soil became recontaminated; this decrease was accompanied by and was generally correlated with an increase in the antibiotic activity of the actinomycetes.

Partial sterilization or "pasteurization" of soil by heat or chemicals can be a biological control method when the pathogen is killed and organisms antagonistic to the pathogen survive the treatment. These native antagonists may then increase to high levels and effectively prevent reestablishment of the pathogen in the soil for a time. *Trichoderma viride* is resistant to partial sterilization of soil by treatments with formalin or carbon disulphide (Evans, 1955; Moubasher, 1963). It also has high recolonizing ability. Bliss (1951) found that *Armillaria mellea*, in citrus root segments buried in soil, was killed when the soil was fumigated with carbon disulfide, but the organism was not killed by soil fumigation when infested root segments were placed in previously sterilized soil.

Trichoderma viride, when placed in heat-sterilized soil, killed the parasite. Bliss postulated that *Trichoderma* was resistant to the fumigant and quickly colonized fumigated soil to a high density level; thus it acted as the killing agent. *Bacillus subtilis,* a heat resistant organism antagonistic to *Rhizoctonia solani,* survived treatment of soil with aerated steam (Olsen and Baker, 1968). Population of the bacterium then increased in steamed soil; this was correlated with decreased survival of *R. solani.*

Changing the pH of soil may affect disease severity through stimulation of antagonists. Recommendations have been made for reducing incidence of potato scab by acidifying soil to 5.2 or lower; growth of *Streptomyces scabies* is inhibited at such low pH levels. Acidifying soil may not always result in simply inhibiting growth of the pathogen. Weindling and Fawcett (1936) controlled damping-off of citrus seedlings, caused by *Rhizoctonia solani,* by acidifying the soil layers next to the seed with aluminum sulfate. There appeared to be an interaction between acidity of the soil and the presence of *Trichoderma lignorum,* since treatment with aluminum sulfate was effective only when *Trichoderma* was present. Heavy inoculation of soil with *T. lignorum* in moderately acid natural soils resulted in a higher percentage of healthy seedlings; this control was absent in neutral soils.

Other practices sometimes influence microbial activity and lead to biological control. For example, Staten and Cole (1948) obtained partial control of *Rhizoctonia solani* damping-off of cotton in areas of low rainfall by irrigating the seed bed 3 weeks before planting. They suggested that the pathogen was partially inhibited by antagonistic or competitive microorganisms whose buildup was encouraged by the preplanting irrigation. They further suggested that partial control might be obtained in areas of abundant rainfall if early seed bed preparation is practiced.

Application of fungicides is usually regarded as "chemical control," but in some cases interactions of these chemicals with the soil microflora occur. In laboratory tests, *Rhizoctonia solani,* growing on PDA with 0.4% pentachloronitrobenzene (PCNB), was killed in less than 10 days after being flooded with a nonsterile water extract of soil. *Rhizoctonia,* on PDA without PCNB, was not killed when flooded with similar extracts. Sterile soil extracts did not have a lethal effect on the organism either in the presence or absence of PCNB (Georgopoulos and Wilhelm, 1962). The authors suggested that either PCNB renders the mycelium of *R. solani* more susceptible to fungicidal factors present in nonsterile soil extracts, or that some antagonistic microorganisms in the extracts are favored. PCNB apparently is decomposed in soil rather quickly, yet its fungicidal action persists for longer periods. Spectrophotometric and gas chromatographic analyses of soil, to which 100 ppm of PCNB had been previously added, revealed that after 166 days 12-33 ppm of PCNB remained, a quantity not sufficient to inhibit *Rhizoctonia* (Ashworth et al., 1965). Severity of *Rhizoctonia* damping-off of bean, however, was substantually reduced in such soil.

Interactions of soil microorganisms with pesticides that are ordinarily considered non-fungicidal may relate to biological control phenomena. Soil-applied herbicides, insecticides, and nematocides or mixtures of these may directly or indirectly affect growth responses of root-infecting fungi and influence disease severity. Some pesticides stimulate growth of antagonistic microorganisms and, at the same time, reduce numbers of plant pathogens (Kaufman, 1964; Curl et al., 1968).

CHAPTER 16
CULTURE MEDIA

A. GENERAL PURPOSE MEDIA

1. *Corn Meal Agar* (Riker and Riker, 1936).

Agar	17 g
Corn meal	20 g
Peptone (if desired)	20 g
Dextrose (if desired)	20 g
Water	1000 ml

Preparation: The corn meal is cooked at about 60 C for one hr in 500 ml of water. The material is filtered through cheesecloth and the filtrate added to melted agar in 500 ml of water. The volume is adjusted before bottling and sterilization.

Use: A common medium for the cultivation of fungi.

2. *Czapek's Sucrose-Nitrate Agar* (Conn, 1921); *Czapek-Dox Agar* (Raper and Thom, 1949).

Agar	15.0 g
$NaNO_3$	2.0 g
K_2HPO_4	1.0 g
$MgSO_4 \cdot 7H_2O$	0.5 g
KCl	0.5 g
$FeSO_4 \cdot 7H_2O$	10.0 mg
Sucrose	30.0 g
Water (distilled)	1000.0 ml

Preparation: The sucrose is added just prior to final sterilization. Czapek-Dox contains 3.0 g of $NaNO_3$ per liter. One g of yeast extract per liter may be added.

Use: A common medium for the cultivation of fungi and bacteria.

3. *Dextrose Agar* (Difco Manual, 1953).

Agar	15 g
Beef extract	3 g
Tryptose	10 g
Dextrose	10 g
NaCl	5 g
Water	1000 ml

Use: For the cultivation of bacteria.

4. *Heart Infusion Agar* (Difco Manual, 1953).

Agar	15 g
Beef heart, infusion from	500 g
Tryptose	10 g
NaCl	5 g
Water (distilled)	1000 ml

Use: A general purpose medium for cultivation of bacteria.

5. *Lima Bean Agar* (Riker and Riker, 1936).

Agar	17 g
Ground lima beans	100 g
Water (distilled)	1000 ml

Preparation: The ground lima beans are soaked in 1000 ml of tepid water for 30 min and then steamed for 30 min. The material is filtered through cheesecloth and the liquid is squeezed out. Agar is added to the liquid after it has been restored to volume with water.

Use: A general purpose medium for cultivation of fungi.

6. *Malt Extract Agar* (Raper and Thom, 1949).

Agar	25 g
Malt extract (Difco)	20 g
Dextrose	20 g
Peptone	1 g
Water (distilled)	1000 ml

Use: A common medium for the cultivation of fungi.

7. *Natural Media* (Snyder and Hansen, 1947).

Agar	15-25 g
Water	1000 ml
Natural material	as desired

Preparation: Agar and water are mixed, autoclaved, and poured into petri dishes. Natural materials, such as pea straw, wheat straw, any above or below ground plant part, or scale or other insects, are chopped into particles of the desired size and sterilized by exposure to propylene oxide in a closed container (Hansen and Snyder, 1947). The fumigant is added to moistened natural material at the rate of about 1 ml per liter capacity of the container. The container lid is loosened after overnight exposure at room temperature to let the fumigant escape, then the material is ready for use. It can be incorporated in melted water agar or placed on the surface of hardened agar in petri dishes. Fumigated natural materials added to sterile water without agar also may be used as a culture medium.

Use: For stimulating growth and/or sporulation of many fungi.

8. *Nutrient Agar* (Difco Manual, 1953).

Agar	15 g
Beef extract	3 g
Peptone	5 g
NaCl	8 g
Water (distilled)	1000 ml

Preparation: The final pH of this medium should be 7.3. Nutrient broth can be prepared by eliminating NaCl and agar from the formula.

Use: For cultivation of bacteria not requiring a highly nutritious medium.

9. *Oatmeal Agar.*

Agar	17 g
Oatmeal	75 g
Yeast extract (if desired)	1 g
Water (distilled)	1000 ml

Preparation: Oatmeal is heated slowly in a water bath on a hot plate for about 1 hr. Agar is melted separately and added to the oatmeal juice after the oatmeal has been strained through cheesecloth. Yeast extract is added and the mixture is sterilized for 20 min at 15 lb. pressure.

Use: A common medium often used for maintaining stock cultures of actinomycetes and fungi.

10. *Potato-Dextrose Agar* (Riker and Riker, 1936).

Agar	17 g
Potatoes (peeled and sliced)	200 g
Dextrose	20 g
Water	1000 ml

Preparation: Potatoes in 500 ml of water are cooked for one hr in a steamer or 40 min in an autoclave. At the same time the agar is melted in 500 ml of water. The potato juice is strained or decanted into the melted agar and the volume is adjusted with water. Dextrose is added before the medium is autoclaved.

Use: A general purpose growth medium for many fungi.

11. *Potato Plugs or Cylinders* (Riker and Riker, 1936).

Preparation: Peeled potatoes are cut with a cork borer or knife so that they will drop freely into culture tubes. Sections are cut about 3-5 cm long, and slanting ¾ of their length. Free starch can be removed by washing cut cylinders in running water for ½ hr. Small wads of absorbent cotton are placed in the bottoms of culture tubes, 2 ml of distilled water are added, and potato cylinders are placed in each tube. The tubes are plugged with cotton and autoclaved.

Use: For cultivation and inducing sporulation of fungi. Form of growth, sporulation, and pigment production on potato plugs are diagnostic characters of species of *Streptomyces*.

12. *Trypticase-Soy Agar* (Rohde, 1968).

Agar	15 g
Trypticase peptone	15 g
Phytone peptone	5 g
NaCl	5 g
Water (distilled)	1000 ml

Use: A general purpose medium for cultivation of bacteria.

13. *Tryptose Agar* (Difco Manual, 1953).

Agar	15 g
Tryptose	20 g
Dextrose	1 g
NaCl	5 g
Thiamine hydrochloride	5 mg
Water (distilled)	1000 ml

Use: A general purpose medium for cultivation of bacteria.

14. *V-8 Juice Agar* (Miller, 1955).

Agar	20 g
V-8 juice (Campbell Soup Co.)	200 ml
$CaCO_3$	3 g
Water (distilled)	800 ml

Use: A general purpose medium for cultivation of fungi.

15. *Wheat Germ-Glucose Agar* (Pisano et al., 1962).

Agar	15 g
Wheat germ	100 g
Glucose	10 g
Water (distilled)	1000 ml

Preparation: Wheat germ is suspended in water which is brought to a boil and simmered for 20 min. The suspension is then centrifuged at 2000 rpm for 20 min. The supernatant liquid is restored to its original volume with water and autoclaved for 15 min at 121 C. Turbidity which may form is removed by filtration through Whatman No. 1 filter paper. Glucose and agar are added to the suspension while boiling. The medium is dispersed in containers and autoclaved.

Use: A general purpose medium for the cutivation of fungi.

B. MEDIA FOR ISOLATING BACTERIA AND ACTINO-MYCETES FROM SOIL

1. *Hutchinson's Agar,* Modified (Bhat and Shetty, 1949).

Agar	20.00 g
Glucose	10.00 g
K_2HPO_4	0.50 g
$MgSO_4 \cdot 7H_2O$	0.20 g
Peptone	0.05 g
KNO_3	0.05 g
Water (distilled)	1000.00 ml

2. *Soil Extract Agar* (Allen, 1957).

Agar	15.0 g
Glucose	1.0 g
K_2HPO_4	0.5 g
Soil extract	100.0 ml
Tap water	900.0 ml
pH	6.8-7.0

Preparation: Soil extract is prepared by heating 1000 g of garden soil with 1000 ml of tap water in an autoclave for 30 min. About 0.5 g of calcium carbonate is added and the soil suspension is filtered through double paper filters until clear. The extract may be bottled and sterilized in 100-ml quantities.

3. *Soil Extract Agar* (Bunt and Rovira, 1955).

Agar	15.00 g
K_2HPO_4	0.40 g
$(NH_4)_2HPO_4$	0.50 g

MgSO$_4$·7H$_2$O	0.05 g
MgCl$_2$	0.10 g
FeCl$_3$	0.01 g
CaCl$_2$	0.10 g
Peptone	1.00 g
Yeast extract	1.00 g
Soil extract	250.00 ml
Tap water	750.00 ml

Preparation: The mixture is adjusted to pH 7.4, steamed for 30 min, and filtered, then autoclaved at 10 lb pressure for 20 min. Soil extract is prepared by autoclaving 1 kg of soil with 1 liter of water for 15 min followed by filtration.

4. *Soil Extract Agar* (James, 1958).

Agar	15.0 g
K$_2$HPO$_4$	0.2 g
Soil extract	1000.0 ml

Preparation: Soil extract is prepared by adding to 500 g of a fertile field soil the amount of water, predetermined for the soil used, necessary to yield 1000 ml of extract. This amount is usually 1200-1500 ml. The soil suspension is heated in an autoclave at 121 C for 30 min. After cooling, it is filtered through paper or cloth. The first cloudy portion of the filtrate is refiltered by returning it to the same filter. Addition of 0.5 g of CaSO$_4$ or CaCO$_3$ will aid in filtering cloudy extracts. To 1000 ml of the extract, K$_2$HPO$_4$ and agar are added. The pH should be adjusted to 6.8.

5. *Soil Extract Agar* (Lochhead, 1940).

Agar	15.0 g
K$_2$HPO$_4$	0.2 g
Soil extract	1000.0 ml

Preparation: Soil extract is prepared by autoclaving 1000 g of soil with 1000 ml of tap water for 20 min at 15 lb. pressure. The material is filtered through paper in a Buchner funnel and made up to 1 liter. The medium is adjusted to pH 6.8 prior to sterilization.

6. *Sodium Albuminate Agar* (Waksman and Fred, 1922).

Agar	15.00 g
Dextrose	1.00 g
K$_2$HPO$_4$	0.50 g
MgSO$_4$·7H$_2$O	0.20 g
Fe$_2$(SO$_4$)$_3$	Trace

Egg albumin	0.25 g
Water (distilled)	1000.00 ml

Preparation: Egg albumin is first dissolved in water and made alkaline to phenolphthalein with 0.1 N NaOH. Final pH of the medium should be 6.8. Egg albumin agar, as described by Waksman (1950), contains 0.15 g of egg albumin.

7. Thornton's Standardized Agar (Thornton, 1922).

Agar	15.0 g
K_2HPO_4	1.0 g
$MgSO_4 \cdot 7H_2O$	0.2 g
$CaCl_2$	0.1 g
NaCl	0.1 g
$FeCl_3$	Trace
KNO_3	0.5 g
Asparagine	0.5 g
Mannitol	1.0 g
Distilled H_2O	1000.0 ml

C. MEDIA FOR ISOLATING ACTINOMYCETES FROM SOIL

1. Arginine-Glycerol-Salt Agar (El-Nakeeb and Lechevalier, 1963).

Agar	15.0 g
Glycerol	12.5 g
Arginine monohydrochloride	1.0 g
K_2HPO_4	1.0 g
NaCl	1.0 g
$MgSO_4 \cdot 7H_2O$	0.5 g
$Fe_2(SO_4)_3 \cdot 6H_2O$	10.0 mg
$CuSO_4 \cdot 5H_2O$	1.0 mg
$ZnSO_4 \cdot 7H_2O$	1.0 mg
$MnSO_4 \cdot H_2O$	1.0 mg
Water (distilled)	1000.0 ml

Preparation: Specific gravity of the glycerol used should be not less than 1.249 at 25 C. Final pH of the medium should be 6.9-7.1.

2. Benedict Agar (Porter et al., 1960).

Agar	20.0 g
Glycerol	20.0 g
L-arginine	2.5 g
NaCl	1.0 g
$FeSO_4 \cdot 7H_2O$	0.1 g

$CaCO_3$	0.1 g
$MgSO_4 \cdot 7H_2O$	0.1 g
Pimaricin	50.0 mg
Water (distilled)	1000.0 ml

3. *Chitin Agar* (Lingappa and Lockwood, 1962).

Agar	20 g
Colloidal chitin	2 g
Water (distilled)	1000 ml

Preparation: Refer to original publication for method of preparing colloidal chitin from crude unbleached or bleached chitin.

4. *Egg Albumin Agar with Actidione* (Corke and Chase, 1956).

Preparation: See ingredients under sodium albuminate agar (Section B). Actidione (40 μg/ml) is added just prior to pouring plates.

5. *Glucose-Asparagine Agar with Sodium Propionate* (Crook et al., 1950).

Agar	15.0 g
Glucose	10.0 g
Sodium propionate	4.0 g
Asparagine	0.5 g
K_2HPO_4	0.5 g
Water (distilled)	1000.0 ml

6. *Glycerol-Asparaginate Agar* (Conn, 1921).

Agar	15 g
Glycerol	10 ml
K_2HPO_4	1 g
Sodium asparaginate	1 g
Water (distilled)	1000 ml

Preparation: Calcium carbonate is added at the rate of 3 g/liter to adjust reaction to approximately pH 7.0.

7. *Jensen's Agar* (Jensen, 1930).

Agar	15.0 g
Dextrose	2.0 g
Casein (dissolved in 10 ml of 0.1 N NaOH)	0.2 g
K_2HPO_4	0.5 g
$MgSO_4 \cdot 7H_2O$	0.2 g

FeCl$_3$·6H$_2$O .... Trace
Water .... 1000.0 ml

Preparation: The mixture should be adjusted to pH 6.5-6.6 prior to sterilization.

8. *Soybean Meal-Glucose Agar* (Tsao et al., 1960).

Agar .... 17.0 g
Soybean meal .... 5.0 g
Glucose .... 5.0 g
CaCO$_3$ .... 0.4 g
Water (distilled) .... 1000.0 ml

Preparation: The ingredients without agar are autoclaved for 20 min. Agar is added to supernatant. Final pH of the medium is adjusted to 7.9-8.1 with 1 N NaOH.

9. *Starch-Casein Agar* (Kuster and Williams, 1964).

Agar .... 18.00 g
Starch .... 10.00 g
Casein (vitamin free) .... 0.30 g
KNO$_3$ .... 2.00 g
NaCl .... 2.00 g
K$_2$HPO$_4$ .... 2.00 g
MgSO$_4$·7H$_2$O .... 0.50 g
CaCO$_3$ .... 0.02 g
FeSO$_4$·7H$_2$O .... 0.01 g
Water (distilled) .... 1000.00 ml

Preparation: For enumeration (or counting) of actinomycete colonies on soil-dilution plates, nystatin and actidione (50 µg/ml of each) should be added to the cooled, liquid medium before pouring into dishes. For purposes of transferring actinomycete colonies from soil-dilution plates, the two antifungal antibiotics plus polymixin B sulfate (5.0 µg/ml) and sodium penicillin (1.0 µg/ml) should be added to the medium (Williams and Davies, 1965). It should be adjusted to pH 7.0-7.2 before autoclaving.

D. MEDIA FOR ISOLATING FUNGI FROM SOIL

1. *Dextrose-Peptone-Yeast Extract Agar* (Papavizas and Davey, 1959).

Agar .... 20.0 g
Dextrose .... 5.0 g

Peptone	1.0 g
Yeast extract	2.0 g
NH_4NO_3	1.0 g
K_2HPO_4	1.0 g
$MgSO_4 \cdot 7H_2O$	0.5 g
$FeCl_3 \cdot 6H_2O$	Trace
Oxgall	5.0 g
Sodium propionate	1.0 g
Water (distilled)	1000.0 ml
Chlortetracycline (Aureomycin)	30.0 mg
Streptomycin	30.0 mg

Preparation: The antibiotics are added to the medium after it has been autoclaved and cooled to 42-45 C.

2. *Ohio Agar* (Schmitthenner and Williams, 1958).

Agar	20.0 g
Glucose	5.0 g
Yeast extract	2.0 g
$NaNO_3$	1.0 g
$MgSO_4 \cdot 7H_2O$	0.5 g
KH_2PO_4	1.0 g
Oxgall	1.0 g
Sodium propionate	1.0 g
Water (distilled)	1000.0 ml
Chloromycetin	50.0 mg
Streptomycin sulfate	50.0 mg

Preparation: Chloromycetin and streptomycin are added after the medium has been autoclaved and cooled to 42-45 C.

3. *Peptone-Dextrose-Rose Bengal Agar* (Martin, 1950).

Agar	20.0 g
KH_2PO_4	1.0 g
$MgSO_4 \cdot 7H_2O$	0.5 g
Peptone	5.0 g
Dextrose	10.0 g
Rose bengal (1%)	3.3 ml
Distilled H_2O	1000.0 ml
Streptomycin	30.0 mg

Preparation: All the materials except rose bengal and streptomycin are dissolved in water. The mixture is heated slowly while stirring until it starts to boil. It is removed from heat and rose bengal is added. After bottling and autoclaving and before pouring plates, streptomycin is added to the cooled liquid medium.

4. *Peptone-Dextrose-Rose Bengal Agar*, Modified (Johnson, 1957).

Preparation: This is Martin's (1950) Peptone-dextrose-rose bengal agar modified by substituting 2 mg/liter of Aureomycin for streptomycin.

5. *Peptone-Dextrose-Phosfon Agar* (Curl, 1968).

Preparation: This is Martin's (1950) Peptone-dextrose-rose bengal agar modified by substituting technical grade Phosfon (plant-growth retardant) at 500 ppm for rose bengal.

6. *Potato-Dextrose Agar with Novobiocin* (Butler and Hine, 1958).

Preparation: Refer to potato-dextrose agar in Section A for ingredients. Novobiocin is heat stable and can be added to the medium (at a final concentration of 100 ppm) before autoclaving.

E. SELECTIVE MEDIA FOR ISOLATING SPECIFIC GROUPS OR GENERA OF MICROORGANISMS FROM SOIL OR PLANTS

1. *Bristol's Sodium Nitrate Solution*, Modified (Bristol, 1920; Allen, 1949).

KH_2PO_4	0.50 g
$NaNO_3$	0.50 g
$MgSO_4 \cdot 7H_2O$	0.15 g
$CaCl_2 \cdot 6H_2O$	0.05 g
NaCl	0.05 g
$FeCl_3 \cdot 6H_2O$	0.01 g
Tap water	1000.00 ml

Use: For isolating soil algae.

2. *Cellulose Agar* (Eggins and Pugh, 1962).

Agar	20.0 g
Cellulose (powdered)	10.0 g
$(NH_4)_2SO_4$	0.5 g
L-asparagine	0.5 g
KH_2PO_4	1.0 g
KCl	0.5 g
$MgSO_4 \cdot 7H_2O$	0.2 g
$CaCl_2$	0.1 g

Yeast extract	0.5 g
Water (distilled)	1000.0 ml

Preparation: Cellulose is prepared as a 4% suspension in water of Whatman's standard grade cellulose powder for chromatography, ball-milled for 72 hr. This suspension of ball-milled cellulose is added to the other constituents of the medium after they have been steamed, and before autoclaving. The final pH of the medium is 6.2.

Use: For isolating cellulose-decomposing fungi from soil.

3. *Gallic Acid Agar* (Flowers and Hendrix, 1969).

Agar	20.0 g
Sucrose	30.0 g
$NaNO_3$	2.0 g
$MgSO_4 \cdot 7H_2O$	0.5 g
KH_2PO_4	1.0 g
Yeast extract	0.5 g
Thiamine hydrochloride	2.0 mg
Gallic acid	425.0 mg
Rose Bengal	0.5 mg
Water (distilled)	1000.0 ml
Pentachloronitrobenzene (PCNB)	25.0 mg
Penicillin G	80,000 units
Nystatin	100,000 units

Preparation: PCNB, Penicillin G, and nystatin are added to the medium after it has been autoclaved and cooled to 42-45 C. The pH is adjusted to 4.5.

Use: For isolating species of *Phytophthora* and *Pythium* from soil.

4. *Kerr's Agar*, Modified (Hendrix and Kuhlman, 1965).

Agar	40.00 g
Sucrose	30.00 g
$NaNO_3$	2.00 g
Yeast extract	0.50 g
$MgSO_4 \cdot 7H_2O$	0.50 g
$FeSO_4$	0.01 g
KH_2PO_4	1.00 g
Streptomycin sulfate	50.00 mg
Rose Bengal	60.00 mg
Water (distilled)	1000.00 ml
PCNB	100.00 mg
Mycostatin	1000 units

Preparation: All ingredients except Mycostatin and PCNB (pentachloronitrobenzene) are combined prior to autoclaving. After the medium has cooled to 42-45 C., and just prior to use, these materials are added. The pH is then adjusted to 4.8 with lactic acid.

Use: For isolating *Phytophthora cinnamomi* and *Pythium* spp. from soil.

5. *Maize Meal Agar* (Duddington, 1955).

Agar	20 g
Maize (corn) meal	20 g
Water	1000 ml

Preparation: The maize meal-water mixture is held at 70 C for one hr. The mixture is cooled and the maize meal settles to the bottom of the container. The supernatant is decanted and filtered through glass wool. The supernatant is then heated and agar is added prior to bottling and autoclaving.

Use: For isolating nematode-trapping fungi from soil.

6. *Mannitol-Nitrate Agar* (Schroth et al., 1965).

Agar	20.0 g
Mannitol	10.0 g
$NaNO_3$	4.0 g
$MgCl_2$	2.0 g
Calcium propionate	1.2 g
Magnesium phosphate	0.2 g
$MgSO_4$	75.0 mg
$NaHCO_3$	75.0 mg
Water (distilled)	1000.0 ml

Preparation: After the medium containing the ingredients listed above has been autoclaved and cooled to 50-55 C, the following materials are added: berberine, 275 mg; sodium selinite, 100 mg; penicillin G (1,625 units/mg), 60 mg; streptomycin sulfate (78.1% streptomycin base), 30 mg; cyclohexamide (85-100%), 250 mg; tyrothricin, 1 mg; and bacitracin (65 units/mg), 100 mg. The pH is adjusted to 7.1 with 1 N HCl.

Use: For isolating *Agrobacterium tumefaciens* from soil.

7. *Patel's Agar,* Modified (Dickey, 1961).

Agar	17 g
Sodium taurocholate	3 g
Peptone	10 g
Dextrose	20 g
Crystal violet (1:1000)	2 ml

Water . . . . 1000 ml
Actidione (85-100% cycloheximide)
(1:1000 dilution) . . . . 100 ml

Preparation: Actidione is added after the medium has been sterilized and cooled to 42-45 C.

Use: For isolating *Agrobacterium tumefaciens* from soil.

8. *PCNB Agar* (Nash and Snyder, 1962).

Agar	20.0 g
Peptone	5.0 g
KH_2PO_4	1.0 g
$MgSO_4 \cdot 7H_2O$	0.5 g
Water (distilled)	1000.0 ml
Streptomycin	300.0 mg
Pentachloronitrobenzene (PCNB), 75% W.P.	1.0 g

Preparation: Streptomycin and PCNB are added after the medium has been autoclaved and cooled to 42-45 C. For routine use the medium need not be autoclaved.

Use: For isolating species of *Fusarium* from soil.

9. *PCNB Agar,* **Modified (Kuhlman, 1966).**

Agar	20.00 g
Peptone	5.00 g
$MgSO_4 \cdot 7H_2O$.	0.25 g
KH_2PO_4	0.50 g
Water (distilled)	1000.00 ml
PCNB (pentachloronitrobenzene)	200 ppm
Penicillin G	50 ppm
Lactic Acid (85%)	1.30 ml
Ethanol	20.00 ml
Sodium desoxycholate	130 ppm

Preparation: Inhibitors and antibiotics are added to the medium after it has been autoclaved and cooled to 42-45 C.

Use: For isolating *Fomes annosus* from soil and from infected woody tissue.

10. *PCNB Agar,* **Modified (Papavizas, 1967).**

Agar	20.0 g

Peptone	5.0 g
KH$_2$PO$_4$	1.0 g
MgSO$_4$·7H$_2$O	0.5 g
Water (distilled)	1000.0 ml
Oxgall	1.0 g
PCNB (active ingredient)	0.5 g
Chlortetracyclene HCl	50.0 mg
Streptomycin Sulfate	100.0 mg

Preparation: Inhibitors and antibiotics are added to the medium after it has been autoclaved and cooled to 42-45 C.

Use: For isolating *Fusarium* from soil.

11. *Reischer's Agar,* Modified (Schmitthenner, 1962).

Agar	20.00 g
KH$_2$PO$_4$	30.00 mg
K$_2$HPO$_4$	30.00 mg
MgSO$_4$·7H$_2$O	20.00 mg
CaCl$_2$	0.56 mg
MnCl$_2$	2.88 mg
ZnCl$_2$	1.67 mg
FeCl$_2$	0.10 mg
Ethylenediaminetetroacetic acid (disodium salt)	11.60 mg
Sucrose	0.41 g
L-asparagine	0.21 g
Thiamine hydrochloride	0.04 mg
Water (distilled)	1000.00 ml
Endomycin	5.00 mg
Streptomycin sulfate	50.00 mg

Preparation: The antibiotics, endomycin and streptomycin, are added to the medium after it has been autoclaved and cooled to 42-45 C.

Use: For isolating species of *Pythium* from soil.

12. *Richard's Agar,* Modified (Martinson and Baker, 1962).

Agar	15.00 g
Sucrose	50.00 g
KNO$_3$	10.00 g
K$_2$HPO$_4$	5.00 g
MgSO$_4$·7H$_2$O	2.50 g

FeCl$_2$	0.02 g
Rose Bengal	67.00 mg
Water (distilled)	1000.00 ml

Use: For isolating *Rhizoctonia solani* from soil in soil microbiology (immersion) sampling tubes.

13. *Rose Bengal Agar*, Modified (Tsao, 1964).

Agar	17.00 g
Dextrose	10.00 g
Yeast extract	0.50 g
KH$_2$PO$_4$	0.50 g
K$_2$HPO$_4$	0.50 g
MgSO$_4$·7H$_2$O	0.50 g
Peptone	0.50 g
Rose Bengal	0.05 g
Water	1000.00 ml
Streptomycin	0.03 g

Preparation: Streptomycin is added to the medium after it has been autoclaved and cooled to 42-45 C.

Use: For isolating *Thielaviopsis basicola* from soil.

14. *Russell's Agar* (Russell, 1956).

Agar	25.0 g
Malt extract	30.0 g
Peptone	5.0 g
O-phenylphenol	0.006 g
Distilled water	1000.0 ml

Preparation: O-phenylphenol is added to the medium after it has been autoclaved and cooled to 42-45 C. The required amount is added from a stock solution prepared by dissolving 1 g in 99 ml of 95% ethyl alcohol.

Use: For isolating Basidiomycetes from infected woody tissue.

15. *Soil Extract Agar*, Dilute (Menzies and Griebel, 1967) (Green and Papavizas, 1968).

Agar	15.0 g
KH$_2$PO$_4$	1.5 g
K$_2$HPO$_4$	4.0 g
Soil extract	25.0 ml
Water (distilled)	975.0 ml
Chlortetracycline	50.0 mg
Chloramphenicol	50.0 mg

Streptomycin	50.0 mg
Polygalacturonic acid	2.0 g

Preparation: Soil extract is prepared by steaming 1 kg of garden soil in 1 liter of tap water for 30 min, then decanting and filtering the suspension. Inhibitors are added after the medium has been autoclaved and cooled to 42-45 C.

Use: For isolating *Verticillium* from soil.

16. Tyrosine-Casein-Nitrate Agar (Menzies and Dade, 1959).

Agar	15 g
Sodium caseinate	25 g
Sodium nitrate	10 g
L-tyrosine	1 g
Water (tap)	1000 ml

Preparation: Sodium caseinate is dissolved in 500 ml of water with gentle heating. The other ingredients are added and the volume is made up to 1 liter with water. The medium is autoclaved for 20 min at 10 lb. pressure.

Use: For isolating *Streptomyces scabies* from soil.

17. VDYA-PCNB Agar (Papavizas, 1964).

Agar	20.0 g
V-8 juice	200.0 ml
CaCO$_3$	1.0 g
Glucose	2.0 g
Yeast extract	2.0 g
Water (distilled)	800.0 ml
PCNB (75% W.P.)	0.66 g
Oxgall	1.00 g
Nystatin	30.00 mg
Streptomycin sulfate	100.00 mg
Chlortetracycline HCl	2.00 mg

Preparation: PCNB, oxgall, and antibiotics are added to the medium after it has been autoclaved and cooled to 42-45 C.

Use: For isolating *Thielaviopsis basicola* from soil.

F. FERMENTATION MEDIA FOR ANTIBIOTIC PRODUCTION

Media used for the production of antibiotic substances can be classified into two groups: synthetic media and complex organic media. Synthetic media usually contain an inorganic source of nitrogen, several salts, and certain supplementary materials such as yeast extract, meat extract, or corn steep. Certain trace elements also may be added. Organic media contain one or more complex forms of nitrogen such as tryptose, peptone, or casein digest, either no other source of carbon is used, or a carbohydrate is added in the form of lactose, glucose, dextrin, starch, brown sugar, molasses, or similar products. If the organism produces much acid, $CaCO_3$ is often added to the medium (Waksman, 1945).

Choice of medium depends on several factors such as nature of the antibiotic substance produced, nature of the organism under investigation, and apparatus at the disposal of the researcher. Since most organisms are selective in producing antibiotics on certain types of media, it is best to vary the ingredients so that several media are tested. Stessel et al. (1953) used five media for screening a number of microorganisms for antibiotic production. Only a few of the media which have been used for antibiotic production are listed here. For further information see Baron (1950).

1. *Actinomycete-Fermentation Medium* (Warren et al., 1955).

Glucose	15.0 g
NaCl	5.0 g
$CaCO_3$	1.0 g
Glycerol	2.5 ml
Soybean meal FF grits (edible grade)	15.0 g
Curbay - B.G.	5.0 g
Water	1000.0 ml

Use: For antibiotic production in shake cultures of many actinomycetes.

2. *Bacitracin-Fermentation Medium* (Anker et al., 1949).

L-glutamic acid	5.00 g
KH_2PO_4	0.50 g
K_2HPO_4	0.50 g
$MgSO_4 \cdot 7H_2O$	0.20 g
$MnSO_4 \cdot 4H_2O$	0.01 g
NaCl	0.01 g
$FeSO_4 \cdot 7H_2O$	0.01 g
$CuSO_4 \cdot 7H_2O$	0.01 g
$CaH_4(PO_4)_2$ (Saturated solution)	2.00 ml
Water	1000.00 ml

Use: For the production of Bacitracin by *Bacillus subtilis* in stationary cultures.

3. *Chloromycetin-Fermentation Medium* (Gottlieb and Diamond, 1951).

Glycerin	10.00 g
Dl-serine	5.00 g
Sodium lactate	11.00 g
$K_2HPO_4 \cdot 3H_2O$	2.39 g
KH_2PO_4	1.39 g
$MgSO_4 \cdot 7H_2O$	1.00 g
NaCl	3.00 g
Water	1000.00 ml

Use: For the production of chloromycetin by *Streptomyces venezuelae*.

4. *Claviformin-Fermentation Medium* (Lochhead et al., 1946).

Glucose	40.00 g
$NaNO_3$	3.00 g
KH_2PO_4	1.00 g
KCl	0.50 g
$MgSO_4 \cdot 7H_2O$	0.50 g
$FeSO_4$	0.01 g
Soil extract	1000.00 ml

Preparation: Soil extract is prepared by autoclaving 1 kg of soil in 1000 ml of water and then filtering.

Use: To produce claviformin from various species of *Penicillium*.

5. *Endomycin-Fermentation Medium* (Gottlieb et al., 1951).

Soybean meal	10.0 g
Cerelose	10.0 g
NaCl	5.0 g
Curbay - B.G.	0.5 g
$CaCO_3$	1.0 g
Water (distilled)	1000.0 ml

Use: For the production of endomycin by *Streptomyces* sp. in shake cultures.

6. *Gliotoxin-Fermentation Medium* (Brian and Hemming, 1945).

Dextrose	25.00 g
Ammonium tartrate	2.00 g
KH_2PO_4	2.00 g

MgSO₄	1.00 g
Ferrous sulfate	0.01 g
Water	1000.00 ml

Use: For the production of gliotoxin by *Trichoderma viride* in stationary cultures.

7. *Test Medium* (Gregory et al., 1952a).

Soya meal	20 g
Glucose	20 g
$CaCO_3$	5 g
Corn steep liquor	5 g
Water (distilled)	1000 ml

Use: To test antibiotic production by a number of Streptomycetes and bacteria in shake cultures.

G. CULTURE SOLUTIONS FOR HIGHER PLANTS

1. *Crone's Solution,* Modified (Rovira, 1956).

K_2HPO_4	1.0 g
KCl	7.0 g
$CaSO_4$	2.5 g
$MgSO_4 \cdot 7H_2O$	2.5 g
$Ca_3(PO_4)_2$	0.5 g
$FePO_4$	0.5 g
KNO_3	2.0 g

Preparation: The salts are mixed and ground. The culture solution is prepared by dissolving 1.5 g of the salt mixture in 1000 ml of water.

2. *Hoagland's Solution I* (Hoagland and Arnon, 1950).

Stock solutions: Molar stock solutions of the following chemicals are made. The quantity of each stock solution in one liter of the final solution is given at the right.

M KH_2PO_4	1 ml
M KNO	5 ml
M $Ca(NO_3)_2$	5 ml
M $MgSO_4$	2 ml

Trace element solution: This solution is prepared by dissolving the indicated quantity of chemicals in 1 liter of water. One ml of this solution is added to each liter of the final solution.

H_3BO_3	2.86 g
$MnCl_2 \cdot 4H_2O$	1.81 g
$ZnSO_4 \cdot 7H_2O$	0.22 g
$CuSO_4 \cdot 5H_2O$	0.08 g
$H_2MoO_4 \cdot H_2O$ (85%)	0.02 g

In addition, iron in the form of 0.5% iron tartrate is added at the rate of 1 ml to each liter of final solution just before it is applied to the plants.

3. *Hoagland's Solution II* (Hoagland and Arnon, 1950).

Stock solutions: Molar stock solutions of the following chemicals are made. The quantity of each stock solution in one liter of the final solution is given at the right.

M $NH_4H_2PO_4$	1 ml
M KNO_3	6 ml
M $Ca(NO_3)_3$	4 ml
M $MgSO_4$	2 ml

A trace element solution and an iron solution are made up and added to the final solution as in Hoagland's Solution I.

4. *Shive and Robbins Solution I* (Shive and Robbins, 1942).

KH_2PO_4	5.9 g
$Ca(NO_3)_2 \cdot 4H_2O$	20.1 g
$MgSO_4 \cdot 7H_2O$	10.7 g
$(NH_4)_2SO_4$	1.8 g
Trace element solution	10.0 ml
Iron solution	100.0 ml
Water	19.0 liters

Preparation: The salts are dissolved separately in 500 ml or more of water to avoid precipitation. The trace element solution is prepared by dissolving together in 500 ml of water, 0.8 g each of boric acid (crystals), manganese sulfate, and zinc sulfate. The iron solution is prepared by dissolving 0.8 g of ferrous sulfate in 500 ml of water containing one drop of concentrated sulfuric acid. The iron solution is added to the final culture solution just before it is applied to the plants.

5. *Shive and Robbins Solution II* (Shive and Robbins, 1942).

KH_2PO_4	3.9 g
$NaNO_3$	6.4 g
$MgSO_4 \cdot 7H_2O$	10.3 g

CaCl$_2$	3.2 g
Trace element solution	10.0 ml
Iron solution	100.0 ml
Water	5.0 gal

Preparation: The materials are prepared as in Shive and Robbins Solution I.

BIBLIOGRAPHY

ABERDEEN, J. E. C. 1955. Quantitative methods for estimating the distribution of soil fungi. Univ. Queensland Papers, Dept. of Botany 3:83-96.
ADAMS, M. H. 1950. Methods of study of bacterial viruses. *In* J. H. Comroe (ed.), Methods in medical research, Vol. 2. The Year Book Publishers, Inc. Chicago, Ill.
ADAMS, P. B. 1967. A buried membrane filter method for studying behavior of soil fungi. Phytopathology 57:602-603.
AGNIHOTHRUDU, V. 1962. A comparison of some techniques for the isolation of fungi from tea soils. Mycopath. et Mycol. Appl. 16:235-242.
ALEXOPOULOS, C. J., and J. A. HERRICK. 1942. Studies in antibiosis between bacteria and fungi. III. Inhibitory action of some actinomycetes on various species of fungi in culture. Bul. Torrey Bot. Club 69:257-261.
ALLEN, O. N. 1949. Experiments in soil bacteriology. 1st ed. Burgess Publ. Co., Minneapolis, Minn.
ALLEN, O. N. 1957. Experiments in soil bacteriology. 3rd rev. ed. Burgess Publ. Co., Minneapolis, Minn.
ALLISON, L. E. 1951. Vapor-phase sterilization of soil with ethylene oxide. Soil Sci. 72:341-352.
ANDAL, R., K. BHUVANESWARI, and N. S. SUBBA-RAO. 1956. Root exudates of paddy. Nature 178:1063.
ANDERSON, A. L. and D. M. HUBER. 1965. The plate-profile technique for isolating soil fungi and studying their activity in the vicinity of roots. Phytopathology 55:592-594.
ANDERSON, E. J. 1951. A simple method for detecting the presence of *Phytophthora cinnamomi* Rands in soil. Phytopathology 41:187-189.
ANKER, H. S., B. A. JOHNSON, J. GOLDBERG, and F. L. MELENEY. 1949. Bacitracin: methods of production, concentration, and partial purification, with a summary of the chemical properties of crude bacitracin. J. Bact. 55:249-255.
ANWAR, A. A. 1949. Factors affecting the survival of *Helminthosporium sativum* and *Fusarium lini* in soil. Phytopathology 39:1005-1019.
ARISTOVSKAYA, T. V. 1962. Principles of ecological analysis in soil microbiology. Pochvovedeniye (A.I.B.S. trans.) 1962(1):4-14.
ARISTOVSKAYA, T. V., and O. M. PARINKINA. 1961. New methods for

studying soil microorganism associations. Pochvovedeniye (A.I.B.S. trans.) 1961(1):12-20.

ARK, P. A., and J. P. THOMPSON. 1961. Detection of hairy root pathogen, *Agrobacterium rhizogenes*, by the use of fleshy roots. Phytopathology 51:69-71.

ASHWORTH, L. J. JR., J. D. PRICE, and B. C. LANGLEY. 1965. A comparisor of biological and chemical assays for pentachloronitrobenezene in a field soil Phytopathology 55:1051.

AUDUS, L. J. 1946. A new soil perfusion apparatus. Nature 158:419.

AYERS, W. A., and R. H. THORNTON. 1968. Exudation of amino acids by intact and damaged roots of wheat and peas. Plant and Soil 28:193-207.

BAKER, K. F. and C. M. OLSEN. 1960. Aerated steam for soil treatment. Phytopathology 50:82.

BAKER, K. F., and C. N. ROISTACHER. 1957. Heat treatment of soil. *In* The U.C. system for producing healthy container-grown plants. California Agr. Exp. Sta. Manual 23:123-196.

BARON, A. L. 1950. Handbook of antibiotics. Reinhold Publ. Co. New York, N. Y.

BARTHA, R., and D. PRAMER. 1965. Features of a flask and method for measuring the persistence and biological effects of pesticides in soil. Soil Sci. 100:68-70.

BHAT, J. V., and M. V. SHETTY. 1949. A suitable medium for the enumeration of the micro-organisms in soil. J. Univ. Bombay. Sect. B, 13:13-15.

BLAIR, I. D. 1942. Studies on the growth in soil and the parasitic action of certain *Rhizoctonia solani* isolates from wheat. Can. J. Res. 20:174-185.

BLAIR, I. D. 1945. Techniques for soil fungus studies. New Zealand J. Sci. Technol. A, 26:258-271.

BLISS, D. E. 1951. The destruction of *Armillaria mellea* in citrus soils. Phytopathology 41:665-683.

BLOOM, J. R., and H. B. COUCH. 1960. Influence of environment on diseases of turfgrasses. I. Effect of nutrition, pH, and soil moisture on *Rhizoctonia* brown patch. Phytopathology 50:532-535.

BOOSALIS, M. G. 1956. Effect of soil temperature and green manure amendments of unsterilized soil on parasitism of *Rhizoctonia solani* by *Penicillium vermiculatum* and *Trichoderma* sp. Phytopathology 46:473-478.

BOOSALIS, M. G., and A. L. SCHAREN. 1959. Methods for microscopic detection of *Aphanomyces euteiches* and *Rhizoctonia solani* and for isolation of *Rhizoctonia solani* associated with plant debris. Phytopathology 49:192-198.

BOOTHROYD, C. W. 1967. Isolation of soil-borne pathogens from soil using plant tissue. *In* Sourcebook of laboratory exercises in plant pathology. W. H. Freeman and Co. San Francisco, Calif.

BOSE, R. G. 1963. A modified cellulosic medium for the isolation of cellulolytic fungi from infected materials and soils. Nature 198:505-506.

BOSWALL, R. L., and D. C. MACKAY. 1963. Vacuum filtration assembly for soil samples. Chem. Anal. 52:54.

BRIAN, P. W. 1957. The ecological significance of antibiotic production. pp. 168-188. *In* Microbial ecology. Cambridge Univ. Press. London.

BRIAN, P. W., and H. G. HEMMING. 1945. Gliotoxin, a fungistatic metabolic product of *Trichoderma viride*. Ann. Appl. Biol. 32:214-220.

BRISTOL, B. M. 1920. On the alga-flora of some desiccated English soils; an important factor in soil biology. Ann. Bot. 34:35-79.

BROADFOOT, W. C. 1933. Studies on foot and root rot of wheat. II. Cultural relationships on solid media of certain microorganisms in association with *Ophiobolus graminis*. Sacc. Can. J. Res. 8(c):545-552.

BROWN, J. C. 1958. Fungal mycelium in dune soils estimated by a modified impression slide technique. Trans. Brit. Mycol. Soc. 41:81-88.

BROWN, J. G., and A. M. BOYLE. 1945. Application of penicillin to crown gall. Phytopathology 35:521-524.

BUCHANAN, R. E., and E. I. FULMER. 1928. Physiology and biochemistry of bacteria. Vol. 1. Williams and Wilkins, Baltimore, Maryland.

BUNT, J. S., and A. D. ROVIRA. 1955. Microbiological studies of some subantarctic soils. J. Soil Sci. 6:119-128.

BURKHOLDER, W. H. 1938. The occurrence in the United States of the tuber ring rot and wilt of the potato. Am. Potato J. 15:243-245.

BUTLER, E. E. 1957. *Rhizoctonia solani* as a parasite of fungi. Mycologia 49:354-373.

BUTLER, E. E., and J. W. ECKERT. 1962. A sensitive method for the isolation of *Geotrichum candidum* from soil. Mycologia 54:106-109.

BUTLER, E. E., and R. B. HINE. 1958. Use of novobiocin for isolation of fungi from soil. Soil Sci. 85:250-254.

BUTLER, F. C. 1953. Saprophytic behavior of some cereal root rot fungi. I. Saprophytic colonization of wheat straw. Ann. Appl. Biol. 40:284-297.

BUXTON, E. W. 1960. Effects of pea root exudate on the antagonism of some rhizosphere microorganisms towards *Fusarium oxysporum* f. *pisi*. Gen. Microbiol. 22:678-689.

BUXTON, E. W. 1962. Root exudates from banana and their relationship to strains of the *Fusarium* causing Panama wilt. Ann. Appl. Biol. 50:269-282.

CALDWELL, R. 1958. Fate of spores of *Trichoderma viride* Pers. ex. Fr. introduced into soil. Nature 181:1144-1145.

CAMPBELL, W. A. 1949. A method of isolating *Phytophthora cinnamomi* directly from soil. Plant Disease Reptr. 33:134-135.

CAMPBELL, W. A., and F. F. HENDRIX, JR. 1967. *Pythium* and *Phytophthora* species in forest soils in the Southeastern United States. Plant Disease Reptr. 51:929-932.

CAMPBELL, W. A., and J. T. PRESLEY. 1945. Design for constant-temperature tanks. Phytopathology 35:213-216.

CARTER, H. P., and J. L. LOCKWOOD. 1957. Methods for estimating numbers of soil microorganisms lytic to fungi. Phytopathology 47:151-154.

CHACKO, C. I., and J. L. LOCKWOOD. 1966. A quantitative method for assaying soil fungistasis. Phytopathology 56:576-577.

CHARPENTIER, M. 1960. Repartition des microorganismes cellulolytiques du sol. I. Les techniques. Ann. Inst. Pasteur 99:153-155.

CHASE, F. E., and P. H. H. GRAY. 1957. Application of the Warburg respirometer in studying respiratory activity in soil. Can. J. Microbiol. 3:335-349.

CHESNIN, L., and W. C. JOHNSON. 1950. Application of centrifugal force to obtain a saturation extract of soil suitable for flame photometric analysis. Soil Sci. 69:497-498.

CHESTERS, C. G. C. 1948. A contribution to the study of fungi in the soil. Trans. Brit. Mycol. Soc. 30:100-117.

CHESTERS, C. G. C., and R. H. THORNTON. 1956. A comparison of techniques for isolating soil fungi. Trans. Brit. Mycol. Soc. 39:301-313.

CHINN, S. H. F. 1953. A slide technique for the study of fungi and actinomycetes in soil with special reference to *Helminthosporium sativum*. Can. J. Bot. 31:718-724.

CHINN, S. H. F., and R. J. LEDINGHAM, and B. J. SALLANS. 1960. Population and viability studies of *Helminthosporium sativum* in field soils. Can. J. Bot. 38:533-539.

CHOLODNY, N. 1930. Über eine neue Methode zur Untersuchung der Bodenmikroflora. Arch. Microb. 1:620-652.

CLARK, F. E. 1949. Soil microorganisms and plant roots. Advances in Agron. 1:241-288.

CONN, H. J. 1918. The microscopic study of bacteria and fungi in the soil. N. Y. Agr. Exp. Sta. Tech. Bul. 64.

CONN, H. J. 1921. The use of various culture media in characterizing actinomycetes. N. Y. Agr. Exp. Sta. Tech. Bul. 83.

COOK, F. D., and A. G. LOCHHEAD. 1959. Growth factor relationships of soil microorganisms as affected by proximity to the plant root. Can. J. Microb. 5:323-334.

COOK, F. D., and C. QUADLING. 1959. A modified technique for isolation of bacteriophage from contaminated materials. Can. J. Microbiol. 5:311-312.
COOK, R. J., ed. 1969. Nature of the influence of crop residues on fungus-induced root diseases. Wash. Agr. Exp. Sta. Bul. 716.
COOK, R. J., and W. C. SNYDER. 1965. Influence of host exudates on growth and survival of germlings of *Fusarium solani* f. *phaseoli* in soil. Phytopathology 55:1021-1025.
COOKE, R. C. 1961. Agar disk method for the direct observation of nematode-trapping fungi in the soil. Nature 191:1411-1412.
COOPER, D. J., K. F. NEILSEN, J. W. WHITE, and W. KALBFLEISCH. 1960. Note on an apparatus for controlling soil temperatures. Can. J. Soil Sci. 40:105-107.
CORKE, C. T., and F. E. CHASE. 1956. The selective enumeration of actinomycetes in the presence of large numbers of fungi. Can. J. Microbiol. 2:12-16.
COUCH, H. B., and J. R. BLOOM. 1960. Influence of soil moisture stresses on the development of the root knot nematode. Phytopathology 50:319-321.
COUCH, H. B., L. H. PURDY, and D. W. HENDERSON. 1967. Application of soil moisture principles to the study of plant disease. Virginia Polytech. Inst., Dept. of Plant Pathology, Bul. 4.
COUCH, J. N. 1939. Technic for collection, isolation and culture of chytrids. J. Elisha Mitchell Sci. Soc. 55:208-214.
CROOK, P., C. C. CARPENTER, and P. F. KLENS. 1950. The use of sodium propionate in isolating actinomycetes from soils. Science 112:656.
CROSSE, J. E. and M. K. HINGORANI. 1958. A method for isolating *Pseudomonas mors-prunorum* phages from the soil. Nature 181:60-61.
CURL, E. A. 1958. Chemical exclusion of mites from laboratory fungal cultures. Plant Disease Reptr. 42:1026-1029.
CURL, E. A. 1961. Influence of sprinkler irrigation and four forage crops on populations of soil microorganisms including those antagonistic to *Sclerotium rolfsii* Sacc. Plant Disease Reptr. 45:517-519.
CURL, E. A. 1963. Control of plant diseases by crop rotation. Bot. Rev. 29:413-479.
CURL, E. A. 1968. Value of a plant-growth retardant for isolating soil fungi. Can. J. Microbiol. 14:182-183.
CURL, E. A., and M. M. ARNOLD. 1964. Influence of substrate and microbial interactions on growth of *Fomes annosus*. Phytopathology 54:1486-1487.
CURL, E. A., and J. D. HANSEN. 1964. The microflora of natural sclerotia of *Sclerotium rolfsii* and some effects upon the pathogen. Plant Disease Reptr. 48:446-450.

CURL, E. A., R. RODRIGUEZ-KABANA, and H. H. FUNDERBURK, JR. 1968. Influence of atrazine and varied carbon and nitrogen amendments on growth of *Sclerotium rolfsii* and *Trichoderma viride* in soil. Phytopathology 58:323-328.

DAMIRGI, S. M., L. R. FREDERICK, and J. M. BREMNER. 1961. Effect of soil dispersion techniques on plate counts of fungi, bacteria, and actinomycetes. Bact. Proc. 61:53.

DAVEY, C. B., and G. C. PAPAVIZAS. 1962. Comparison of methods for isolating *Rhizoctonia* from soil. Can. J. Microbiol. 8:847-853.

DAVEY, C. B., and S. A. WILDE. 1955. Determination of the numbers of soil microorganisms by the use of molecular membrane filters. Ecology 36:760-761.

DAVIES, B. E., and R. I. DAVIES. 1963. A simple centrifugation method for obtaining small samples of soil solution. Nature 198:216-217.

DAVIES, F. R. 1935. Superiority of silver nitrate over mercuric chloride for surface sterilization in the isolation of *Ophiobolus graminis* Sacc. Can. J. Res. 13(c):168-173.

DEEMS, R. E., and H. C. YOUNG. 1956. Black root of sugar beets as influenced by various cropping sequences and their associated mycofloras. Phytopathology 46:10.

DICK, M. W., and H. V. NEWBY. 1961. The occurrence and distribution of Saprolegniaceae in certain soils of southeast England. I. Occurrence. J. Ecol. 49:403-419.

DICKEY, R. S. 1961. Relation of some edaphic factors to *Agrobacterium tumefaciens*. Phytopathology 51:607-614.

DIFCO MANUAL, 9th ed. 1953. Difco Laboratories, Inc., Detroit, Mich.

DIMOCK, A. W. 1967. A soil-temperature control system employing air as the heat-transfer medium. Plant Disease Reptr. 51:873-876.

DOBBS, C. G., and W. H. HINSON. 1953. A widespread fungistasis in soils. Nature 172:197-199.

DRECHSLER, C. 1929. The beet water mold and several related root parasites. J. Agr. Res. 38:309-361.

DRECHSLER, C. 1938. Two hypomycetes parasitic on oospores of root-rotting oomycetes. Phytopathology 28:81-103.

DUDDINGTON, C. L. 1955. Notes on the technique of handling predacious fungi. Trans. Brit. Mycol. Soc. 38:97-103.

DUKES, P. D., and J. L. APPLE. 1965. Effect of oxygen and carbon dioxide tensions on growth and inoculum potential of *Phytophthora parasitica* var. nicotianae. Phytopathology 55:666-669.

ECKERT, J. W., and P. H. TSAO. 1962. A selective antibiotic medium for isolation of *Phytophthora* and *Pythium* from plant roots. Phytopathology 52:771-777.

EDWARD, J. C., and R. D. DURBIN. 1959. Inhibition of fungi by *Acrophialophora nainiana.* Plant Disease Reptr. 43:1191-1194.

EGDELL, J. W. et al. 1960. Some studies of the colony count technique for soil bacteria. J. Appl. Bact. 23:69-86.

EGGINS, H. O. W., and G. J. F. PUGH. 1962. Isolation of cellulose-decomposing fungi from the soil. Nature 193:94-95.

ELKAN, G.H., and W. E. C. MOORE. 1962. A rapid method for measurement of CO_2 evolution by soil microorganisms. Ecology 43:775-776.

EL-NAKEEB, M. A., and H. A. LECHEVALIER. 1963. Selective isolation of aerobic actinomycetes. Appl. Microbiol. 11:75-77.

EREN, J., and D. PRAMER. 1965. The most probable number of nematode-trapping fungi in soil. Soil Sci. 99:285.

EREN, J., and D. PRAMER. 1966. Application of immunofluorescent staining to studies of the ecology of soil microorganisms. Soil Sci. 101:39-45.

ERWIN, D. C. 1954. Root rot of alfalfa caused by *Phytophthora cryptogea.* Phytopathology 44:700-704.

EVANS, E. 1955. Survival and recolonization by fungi in soil treated with formalin or carbon disulphide. Trans. Brit. Mycol. Soc. 38:335-346.

EVANS, E., and D. GOTTLIEB. 1955. Gliotoxin in soils. Soil Sci. 80:295-301.

EVANS, G., W. C. SNYDER, and S. WILHELM. 1966. Inoculum increase of the Verticillium wilt fungus in cotton. Phytopathology 56:590-594.

EVANS, G., S. WILHELM, and W. C. SNYDER. 1967. Quantitative studies by plate counts of propagules of the Verticillium wilt fungus in cotton field soils. Phytopathology 57:1250-1255.

FARLEY, J. D. 1967. The suppression of actinomycetes by PCNB in culture media used for enumerating soil bacteria. Phytopathology 57:811.

FAUST, M. A. and D. PRAMER. 1964. A staining technique for the examination of nematode-trapping fungi. Nature 204:94-95.

FERGUSON, J. 1953. Factors in colonization of sclerotia by soil organisms. Phytopathology 43:471.

FERGUSON, J. 1957. Personal communication.

FERRIS, J. M., B. LEAR, A. W. DIMOCK, and W. F. MAI. 1955. A description of Cornell temperature tanks. Plant Disease Reptr. 39:875-878.

FLENTJE, N. T., R. L. DODMAN, and A. KERR. 1963. The mechanism of host penetration by *Thanatephorus cucumeris.* Australian J. Biol. Sci. 16:784-799.

FLINT, L. H. 1947. Antibiotic activity in the genus *Haplosiphon.* La. Acad. Sci., Proc. 10:30-31.

FLOWERS, R. A., and J. W. HENDRIX. 1969. Gallic acid in a procedure for isolation of *Phytophthora parasitica* var. *nicotianae* and *Pythium* spp. from soil. Phytopathology 59:725-731.
FREEMAN, T. E., and E. C. TIMS. 1955. Antibiosis in relation to pink root of shallots. Phytopathology 45:440-442.
GABE, D. R. 1961. Capillary method for studying microbe distribution in soils. Pochvovedeniye (A.I.B.S. trans.) 1961(1):70-75.
GAMBLE, S. J. R., T. W. EDMINSTER, and F. S. ORCUTT. 1952. Influence of double-cut plow mulch tillage on number and activity of microorganisms. Soil Sci. Soc. Amer., Proc. 16:267-269.
GAMS, W. 1969. Isolierung von Hyphen aus dem Boden. Sydowia 13:87-94.
GARRETT, S. D. 1938. Soil conditions and the take-all disease of wheat. III. Decomposition of the resting mycelium of *Ophiobolus graminis* in infected wheat stubble buried in the soil. Ann. Appl. Biol. 25:742-766.
GARRETT, S. D. 1956a. Biology of root-infecting fungi. Cambridge University Press, England.
GARRETT, S. D. 1956b. Rhizomorph behavior in *Armillaria mellea* (Vahl) Quel. II. Logistics of infection. Ann. Bot (n. s.) 20:193-209.
GEORGOPOULOS, S. G., and S. WILHELM. 1962. Effect of nonsterile soil on Rhizoctonia solani mycelium in the presence of PCNB. Phytopathology 52:361.
GHABRIAL, S. A., and T. P. PIRONE. 1966. Effect of mineral nutrition on host reaction to *Cladosporium fulvum.* Phytopathology 56:493-496.
GILBERT, W. W. 1926. An improved method for isolation of *Thielavia basicola.* Phytopathology 16:579.
GILMORE, A. E. 1959. A soil sampling tube for soil microbiology. Soil Sci. 87:95-99.
GILMOUR, C. M., L. DAMSKY, and W. B. BOLLEN. 1958. Manometric gas analysis as an index of microbial oxidation and reductions in soil. Can. J. Mircobiol. 4:287-293.
GOLEBIOWSKA, J. 1957. L'influence du mode de prélèvement et du conservation des échantillons du sol sur son état microbiologique. Pedologie (No. spec.) 7:98-103.
GOODMAN, J. J., and A. W. HENRY. 1947. Action of subtilin in reducing infection by a seed-borne pathogen. Science 105:320-321.
GOOS, R. D. 1960. Basidiomycetes isolated from soil. Mycologia 52:661-663.
GOPALKRISHNAN, K. S., and J. A. JUMP. 1952. The antibiotic activity of thiolutin in the chemotherapy of the Fusarium wilt of tomato. Phytopathology 42:338-339.
GOTH, R. W., J. E. DEVAY, and F. J. SCHICK. 1967. A quantitative method for the isolation of *Pythium* species from soil using sweet corn. Phytopathology 57:813.

GOTTLIEB, D., P. K. BHATTACHARYYA, H. E. CARTER, and H. W. ANDERSON. 1951. Endomycin, a new antibiotic. Phytopathology 41:393-400.
GOTTLIEB, D., and L. DIAMOND. 1951. A synthetic medium for chloromycetin. Bul. Torrey Bot. Club 78:56-60.
GOTTLIEB, D., and P. SIMINOFF. 1952. The production and role of antibiotics in the soil. II. Chloromycetin. Phytopathology 42:91-97.
GREEN, R. J., and G. C. PAPAVIZAS. 1968. The effect of carbon source, carbon to nitrogen ratios, and organic amendments on survival of propagules of *Verticillium albo-atrum* in soil. Phytopathology 58:567-570.
GREGORY, K. F., O. N. ALLEN, A. J. RIKER, and W. H. PETERSON. 1952a. Antibiotics as agents for the control of certain damping-off fungi. Amer. J. Bot. 39:405-415.
GREGORY, K. F., O. N. ALLEN, A. J. RIKER, and W. H. PETERSON. 1952b. Antibiotics and antagonistic microorganisms as control agents against damping-off of alfalfa. Phytopathology 42:613-622.
GROSSBARD, E. 1952. Antibiotic production by fungi on organic manures and in soil. J. Gen. Microbiol. 6:295-310.
GROSSBARD, E., and D. M. HALL. 1964. An investigation into the possible changes in the microbial population of soils stored at $-15°C$. Plant and Soil 21:317-332.
HAAS, J. H. 1964. Isolation of *Phytophthora megasperma* var. *sojae* in soil dilution plates. Phytopathology 54:894.
HACK, J. E. 1957. The effect of spores germination and development on plate counts of fungi in soil. J. Gen. Microbiol. 17:625-630.
HALVORSON, H. O., and N. R. ZIEGLER. 1933. Quantitative bacteriology. Burgess Publ. Co., Minneapolis, Minn.
HAMS, A. F., and G. D. WILKIN. 1961. Observations on the use of predacious fungi for the control of *Heterodera* spp. Ann. Appl. Biol. 49:515-523.
HANSEN, H. N., and W. C. SNYDER. 1947. Gaseous sterilization of biological materials for use as culture media. Phytopathology 37:369-371.
HANSON, A. M. 1945. A morphological, developmental, and cytological study of four saprophytic chytrids. I. *Catenomyces persicinus* Hanson. Amer. J. Bot. 32:431-438.
HARLEY, J. L., and J. S,WAID. 1955. A method of studying active mycelia on living roots and other surfaces in the soil. Trans. Brit. Mycol. Soc. 38:104-118.
HARRIS, P. J. 1969. Errors in direct counts of soil organisms due to bacteria in agar powders. Soil Biol. Biochem. 1:103-104.
HARRISON,M. D., and C. H. LIVINGSTON. 1966. A method for isolating *Verticillium* from field soil. Plant Disease Reptr. 50:897-899.
HARRISON, M. D., C. H. LIVINGSTON, and NAGAYOSHI OSHIMA. 1965.

An improved system for controlling soil temperatures in the study of soil-borne plant pathogens. Plant Disease Reptr. 49:452-454.
HARRISON, R. W. 1955. A method of isolating vesicular arbuscular endophytes from roots. Nature 175:432.
HARVEY, J. V. 1925. A survey of the water molds and Pythiums occurring in the soils of Chapel Hill. J. Elisha Mitchell Sci. Soc. 41:151-164.
HEALY, M. J. R. 1962. Some basic statistical techniques in soil zoology. In Progress in soil zoology, pp. 3-9, Butterworths, London.
HENDERSON, M. E. K. 1961. Isolation, identification and growth of some soil hyphomycetes and yeast-like fungi which utilize aromatic compounds related to lignin. J. Gen. Microbiol. 26:149-154.
HENDRIX, F. F., JR., and E. G. KUHLMAN. 1962. A comparison of isolation methods for *Fomes annosus*. Plant Disease Reptr. 46:674-676.
HENDRIX, F. F., JR., and E. G. KUHLMAN. 1965. Factors affecting direct recovery of *Phytophthora cinnamomi* from soil. Phytopathology 55:1183-1187.
HENIS, Y., and I. CHET. 1968. The effect of nitrogenous amendments on the germinability of sclerotia of *Sclerotium rolfsii* and on their accompanying microflora. Phytopathology 58:209-211.
HERR, L. J. 1959. A method of assaying soils for numbers of actinomycetes antagonistic to fungal pathogens. Phytopathology 49:270-273.
HESSAYON, D. G. 1953. Fungitoxins in the soil. II. Trichothecin, its production and inactivation in unsterilized soils. Soil Sci. 75:395-404.
HESSELTINE, C. W., M. HAUCK, M. T. HAGEN, and N. BOHONOS. 1952. Isolation and growth of yeasts in the presence of Aureomycin. J. Bact. 64:55-61.
HILDEBRAND, A. A., and P. M. WEST. 1941. Strawberry root rot in relation to microbiological changes induced in root rot soil by the incorporation of certain cover crops. Can. J. Res. 19(c):183-198.
HINE, R. B., and L. V. LUNA. 1963. A technique for isolating *Pythium aphanidermatum* from soil. Phytopathology 53:727-728.
HINE, R. B., and E. E. TRUJILLO. 1966. Manometric studies on residue colonization in soil by *Pythium aphanidermatum* and *Phytophthora parasitica*. Phytopathology 56:334-336.
HIRTE, W. 1962. Einige Untersuchungen zur Methodik der mikrobiologischen Bodenprobenverarbeitung. Zentr. f. Bakteriol, II Abt. 115:394-403.
HOAGLAND, D. R., and D. I. ARNON. 1950. The water-culture method for growing plants without soil. California Agr. Exp. Sta. Circ. 347.
HOFMANN, E. 1963. Die Analyse von Enzymen im Boden, p. 416-423. In Moderne Methoden der Pflanzenanalyse. Vol. VI. Springer-Verlag, Berlin.

HORNBY, D., and A. J. ULLSTRUP. 1965. Physical problems of sampling soil suspensions in the dilution-plate technique. Phytopathology 55:1062.

HSU, S. C., and J. L. LOCKWOOD. 1969. Nutrient deprivation as a mechanism of inhibition of fungi in agar by streptomycetes. Phytopathology 59:1032.

HUBERT, E. E. 1953. Aids in isolation of fungi from woody tissues. Phytopathology 43:403-404.

ISAAC, I. 1954. Studies in the antagonism between *Blastomyces luteus* and species of *Verticillium*. Ann. Appl. Biol. 41:305-310.

JACKSON, M. L. 1958. Soil chemical analysis. Prentice-Hall, Inc., Englewood Cliffs, N. J.

JACKSON, R. M. 1958. An investigation of fungistasis in Nigerian soils. J. Gen. Microbiol. 18:248-258.

JAMES, N. 1958. Soil extract in soil microbiology. Can. J. Microbiol. 4:363-370.

JAMES, N. 1959. Plate counts of bacteria and fungi in a saline soil. Can. J. Microbiol. 5:431-439.

JENSEN, H. L. 1930. Actinomycetes in Danish soils. Soil Sci. 30:59-77.

JENSEN, V. 1962. Studies on the microflora of Danish beech forest soils. I. The dilution plate count technique for the enumeration of bacteria and fungi in soil. Zentralblatt f. Bakteriologie, etc. II. Abt. Bd. 116:13-32.

JOHANSEN, D. A. 1940. Plant Microtechnique. McGraw Hill, New York.

JOHN, R. P. 1942. An ecological and taxonomic study of the algae of British soils. I. The distribution of surface growing algae. Ann. Bot. N. S. 6:323-349.

JOHNSON, H. G. 1957. A method for determining the degree of infestation by pea root-rot organisms in soil. Phytopathology 47:18.

JOHNSON, L. F. 1952. The relation of antagonistic microorganisms to Pythium root rot of sugarcane and corn in recontaminated soils. Proc. La. Acad. Sci. 15:24-31.

JOHNSON, L. F. 1953. The effect of antagonistic soil microorganisms on the severity of Pythium root rot of sugarcane. Dissertation. Louisiana State University, Baton Rouge, La.

JOHNSON, L. F. 1954. Antibiosis in relation to Pythium root rot of sugarcane and corn. Phytopathology 44:69-73.

JOHNSON, L. F. 1957. Effect of antibiotics on the numbers of bacteria and fungi isolated from soil by the dilution-plate method. Phytopathology 47:630-631.

JOHNSON, L. F. 1962. Effect of the addition of organic amendments to soil on root knot of tomatoes. II. Relation of soil temperature, moisture, and pH. Phytopathology 52:410-413.

JOHNSON, L. F. 1964. Survival of fungi in soil exposed to gamma radiation. Can. J. Bot. 42:105-113.
JOHNSON, L. F., A. Y. CHAMBERS, and J. W. MEASELLS. 1969. Influence of soil moisture, temperature, and planting date on severity of cotton seedling blight. Tenn. Agr. Exp. Sta. Bul. 461.
JOHNSON, L. F., E. A. CURL, J. H. BOND, and H. A. FRIBOURG. 1959. Methods for studying soil microflora-plant disease relationships. Burgess Publ. Co., Minneapolis, Minn.
JOHNSON, L. F., and K. MANKA. 1961. A modification of Warcup's soil-plate method for isolating soil fungi. Soil Sci. 92:79-84.
JONES, P. C. T., and J. E. MOLLISON. 1948. A technique for the quantitative estimation of soil microorganisms. J. Gen. Microbiol. 2:54-69.
JOSLYN, D. A., and M. GALBRAITH. 1950. A turbidimetric method for the assay of antibiotics. J. Bact. 59:711-716.
KATZNELSON, H. 1946. The "rhizosphere effect" of mangels on certain groups of soil microorganisms. Soil Sci. 62:343-354.
KATZNELSON, H., J. W. ROUATT, and T. M. B. PAYNE. 1955. The liberation of amino acids and reducing compounds by plant roots. Plant and Soil 7:35-48.
KAUFMAN, D. D. 1964. Effect of s-triazine and phenylurea herbicides on soil fungi in corn- and soybean-cropped soil. Phytopathology 54:897.
KAUFMAN, D. D. 1966. An inexpensive, positive pressure, soil perfusion system. Weeds 14:90-91.
KAUFMAN, D. D., and L. E. WILLIAMS. 1962. Polyethylene bags for the study of soil microorganisms. Phytopathology 52:16.
KELNER, A. 1948. A method for investigating large microbial populations for antibiotic activity. J. Bact. 56:157-162.
KERR, A. 1953. A method of isolating soft-rotting bacteria from soils. Nature 172:1155.
KERR, A. 1956. Some interactions between plant roots and pathogenic soil fungi. Aust. J. Biol. Sci. 9:45-52.
KING, C. J., and C. HOPE. 1932. Distribution of the cotton root-rot fungus in soil and in plant tissues in relation to control by disinfectants. J. Agr. Res. 45:725-740.
KING, C. J., C. HOPE, and E. D. EATON. 1934. Some microbiological activities affected in manurial control of cotton root rot. J. Agr. Res. 49:1093-1107.
KITZKE, E. D. 1952. A new method for isolating members of the Acrasieae from soil samples. Nature 170:284-285.
KLECZKOWSKA, J. 1945. The production of plaques by Rhizobium bacteriophage in poured plates and its value as a counting method. J. Bact. 50:71-79.

KLEMMER, H. W., and R. Y. NAKANO. 1962. Techniques in isolation of Pythiaceous fungi from soil and diseased pineapple tissue. Phytopathology 52:955-956.

KLEMMER, H. W., and R. Y. NAKANO. 1964. A semi-quantitative method of counting nematode-trapping fungi in soil. Nature 203:1085.

KLOTZ, L. J., and T. A. DEWOLFE. 1958. Techniques for isolating *Phytophthora* spp. which attack citrus. Plant Disease Reptr. 42:675-676.

KOIKE, H. 1967. An agar-layer method useful in detecting antibiotic-producing microorganisms against *Pythium graminicola.* Plant Disease Reptr. 51:333-335.

KOIKE, H., and P. L. GAINEY. 1958. Studies on antibiotic production in soil: I. Use of fritted-filter tubes for testing antibiotics in solution. Soil Sci. 86:98-102.

KOMMEDAHL, T., and I. PIN CHANG. 1966. Coating corn kernels with microorganisms to control seedling blight caused by *Fusarium roseum.* Phytopathology 56:885.

KRUGER, W. 1959. The control of tomato canker [*Corynebacterium michiganense* (Erw. Smith) Jensen] by means of antibiotics. S. African J. Agr. Sci. 2:195-205.

KRUGER, W. 1960. Antibiotics as seed protectants (1958/59). S. African J. Agr. Sci. 3:409-418.

KRUGER, W. 1961. The activity of antibiotics in soil. I. Adsorption of antibiotics by soils. S. African J. Agr. Sci. 4:171-183.

KRUGER, W. 1961. The activity of antibiotics in soil. II. Movement, stability, and biological activity of antibiotics in soils and their uptake by tomato plants. S. African J. Agr. Sci. 4:301-313.

KUHLMAN, E. G. 1966. Recovery of *Fomes annosus* spores from soil. Phytopathology 56:885.

KUMAR, D., and R. F. PATTON. 1964. Fluorescent antibody technique for detection of *Polyporus tomentosus.* Phytopathology 54:898.

KUSTER, E., and S. T. WILLIAMS. 1964. Selection of media for isolation of Streptomycetes. Nature 202:928-929.

LAWRENCE, C. H. 1956. A method of isolating actinomycetes from scabby potato tissue and soil with minimal contamination. Can. J. Bot. 34:44-47.

LEACH, L. D., and A. E. DAVEY. 1938. Determining the sclerotial population of *Sclerotium rolfsii* by soil analysis and predicting losses of sugar beets on the basis of these analyses. J. Agr. Res. 56:619-631.

LEDINGHAM, R. J., and S. H. F. CHINN. 1955. A flotation method for obtaining spores of *Helminthosporium sativum* from soil. Can. J. Bot. 33:298-303.

LEDINGHAM, R. J., B. J. SALLANS, and P. M. SIMMONDS. 1949. The significance of the bacterial flora on wheat seed in inoculation studies with *Helminthosporium sativum.* Sci. Agr. 29:253-262.

LEGGE, B. J. 1952. Use of glass fibre material in soil mycology. Nature 169:759-760.

LEHNER, A., W. NOWAK, and L. SEIHOLD. 1958. Eine Weiterentwicklung des Boden-Fluorochomierungs-Verfahrens mit Acridinorange zur Kombination methode. Landw. forsch. 11:121-127.

LENHARD, G. 1956. Die Dehydrogenaseaktivität des Bodens als Mass fur die Mikroorganismentätigkeit im Boden. Z. Pflanzenernahr., Dung., Bodenk. 73:1-11.

LENHARD, G. 1957. Die Dehydrogenaseaktivität des Bodens als Mass für die Menge an mikrobioell abbaubaren Humussoffen. Z. Pflanzenernahr., Dung., Bodenk. 77:193-198.

LEO, M. W. M. 1963. A new suction plate apparatus for extraction of soil solution in conductivity determination. Soil Sci. 95:142-143.

LEVISOHN, I. 1955. Isolation of ectotrophic mycorrhizal mycelia from rhizomorphs present in soil. Nature 176:519.

LINGAPPA, Y., and J. L. LOCKWOOD. 1962. Chitin media for selective isolation and culture of actinomycetes. Phytopathology 52:317-323.

LINGAPPA, B. T., and J. L. LOCKWOOD. 1963. Direct assay of soil for fungistasis. Phytopathology 53:529-531.

LIPMAN, C. B., and D. E. MARTIN. 1918. Are unusual precautions necessary in taking soil samples for ordinary bacteriological tests? Soil Sci. 6:131-136.

LIU, S., and E. K. VAUGHAN. 1965. Control of Pythium infection in table beet seedlings by antagonistic microorganisms. Phytopathology 55:986-989.

LLOYD, A. B., and J. L. LOCKWOOD. 1962. Precautions in isolating *Thielaviopsis basicola* with carrot discs. Phytopathology 52:1314-1315.

LOCHHEAD, A. G. 1940. Qualitative studies of soil microorganisms. III. Influence of plant growth on the character of the bacterial flora. Can. J. Res. 18(c):42-53.

LOCHHEAD, A. G., F. E. CHASE, and G. B. LANDERKIN. 1946. Production of claviformin by soil Penicillia. Can. J. Res. 24(e):1-9.

LOCHHEAD, A. G., and G. B. LANDERKIN. 1949. Aspects of antagonisms between microorganisms in soil. Plant and Soil 1:271-276.

LOO, Y. H. et al. 1945. Assay of streptomycin by the paper-disc plate method. J. Bact. 50:701-709.

LOUW, H. A., and D. M. WEBLEY. 1959. The bacteriology of the root region of the oat plant grown under controlled pot culture conditions. J. Appl. Bact. 22:216-226.

LUCAS, R. L. 1955. A comparative study of *Ophiobolus graminis* and *Fusarium culmorum* in saprophytic colonization of wheat straw. Ann. Appl. Biol. 43:134-143.

LUDWIG, R. A., and A. W. HENRY. 1943. Studies on the microbiology of recontaminated sterilized soil in relation to its infestation with *Ophiobolus graminis* Sacc. Can. J. Res. 21(c):343-350.
LUTTRELL, E. S. 1967. A strip bait for studying the growth of fungi in soil and aerial habitats. Phytopathology 57:1266-1267.
MACWITHEY, H. S. 1957. Another modification of the Chester tube method for. examination of soil microflora. Report of the 36th Ann. Conv. of Northwest Assn. of Horticulturists, Entomologists, and Plant Pathologists, pp. 5-6.
MAIER, C. R. 1959. Effect of certain crop residues on bean root-rot pathogens. Plant Disease Reptr. 43:1027-1030.
MAKKONEN, R., and O. POHJAKALLIO. 1960. On the parasites attacking the sclerotia of some fungi pathogenic to higher plants and on the resistance of these sclerotia to their parasites. Acta Agric. Scand. 10:105-126.
MALOY, O. C., and M. ALEXANDER. 1958. The "most probable number" method for estimating populations of plant pathogenic organisms in the soil. Phytopathology 48:126-128.
MANKA, K. 1964. Proby dalszego udoskonalenia zmodyfikowanej metody warcupa izolowania grzybow z gleby. Poznanskie Towarzystwo Przyjaciol Nauk 17:29-45.
MANKA, K., A. BLONSKA, and S. WNEKOWSKI. 1961. Badania nad skladem mikroflory kilku rodzajow gleb i jej oddzialywaniem na rozwoj niektorych pasozytnicqych grzybow glebowych. Prace Nauk. Inst. Ochr. Roslin, Poznan. 3:145-231.
MANKAU, R. 1961. An attempt to control root-knot nematode with *Dactylaria thaumasia* and *Arthrobotrys arthrobotryoides*. Plant Disease Reptr. 45:164-166.
MANKAU, S. K., and R. MANKAU. 1962. Multiplication of *Aphelenchus avenae* on phytopathogenic soil fungi. Phytopathology 52:741.
MARTIN, J. P. 1950. Use of acid, rose bengal and streptomycin in the plate method for estimating soil fungi. Soil Sci. 69:215-232.
MARTINSON, C. A. 1963. Inoculum potential relationships of *Rhizoctonia solani* measured with soil microbiological sampling tubes. Phytopathology 53:634-638.
MARTINSON, C., and R. BAKER. 1962. Increasing relative frequency of specific fungus isolations with soil microbiological sampling tubes. Phytopathology 52:619-621.
MATTURI, S. T., and H. STENTON. 1958 A technique for the investigation of the competitive saprophytic ability of soil fungi by the use of easily decomposed substrates. Nature 182:1248-1249.
McCAIN, A. H. 1967. Quantitative recovery of sclerotia of *Sclerotium cepivorum* from field soil. Phytopathology 57:1007.

McCAIN, A. H., O. V. HOLTZMANN, and E. E. TRUJILLO. 1967. Concentration of *Phytophthora cinnamomi* chlamydospores by soil sieving. Phytopathology 57:1134-1135.
McDANIEL, H. R., J. B. MIDDLEBROOK, and R. O. BOWMAN. 1962. Isolation of pure cultures of algae from contaminated cultures. Appl. Microbiol. 10:223.
McKEE, R. K., and A. E. W. BOYD. 1952. Dry-rot disease of the potato. III. A biological method of assessing soil infectivity. Ann. Appl. Biol. 39:44-53.
McKEEN, W. E. 1949. A study of sugar beet root rot in southern Ontario. Can. J. Res. 27(c):284-311.
McKEEN, W. E. 1958. Red stele root disease of the loganberry and strawberry caused by *Phytophthora fragariae*. Phytopathology 48:129-132.
McLAREN, A. D., R. A. LUSE, and J. J. SKUJINS. 1962. Sterilization of soil by irradiation and some further observations on soil enzyme activity. Soil Sci. Soc. Am., Proc. 26:371-377.
MEAD, H. W. 1933. Studies of methods for the isolation of fungi from wheat roots and kernels. Sci. Agric. 13:304-312.
MENNA, M. E. di. 1957. The isolation of yeasts from soil. J. Gen. Microbiol. 17:678-688.
MENZIES, J. D. 1963. The direct assay of plant pathogen populations in soil. Ann. Rev. Phytopathology 1:127-142.
MENZIES, J. D., and C. E. DADE. 1959. A selective indicator medium for isolating *Streptomyces scabies* from potato tubers or soil. Phytopathology 49:457-458.
MENZIES, J. D., and G. E. GRIEBEL. 1967. Survival and saprophytic growth of *Verticillium dahliae* in uncropped soil. Phytopathology 57:703-709.
MERIDITH, C. H. 1944. The antagonism of soil organisms to *Fusarium oxysporum cubense*. Phytopathology 34:426-429.
MILLER, J. J., and N. S. WEBB. 1954. Isolation of yeasts from soil with the aid of acid, rose bengal, and oxgall. Soil Sci. 77:197-204.
MILLER, P. M. 1955. V-8 juice agar as a general-purpose medium for fungi and bacteria. Phytopathology 45:461-462.
MINDERMAN, G. 1956. The preparation of microtome sections of unaltered soil for the study of soil organisms *in situ*. Plant and Soil 8:42-48.
MIRCHINK, T. G., and K. P. GRESHNYKH. 1962. Toxin formation in the soil by some fungal species of the genus *Penicillium*. Microbiol. (English trans. of Mikrobiologiya) 30:851-854.
MITCHELL, R. 1963. Addition of fungal cell-wall components to soil for biological disease control. Phytopathology 53:1068-1071.
MITCHELL, R., and M. ALEXANDER. 1961. Chitin and the biological control of Fusarium diseases. Plant Disease Reptr. 45:487-490.

MITCHELL, R., and M. ALEXANDER. 1962. Microbiological processes associated with the use of chitin for biological control. Soil Sci. Soc. Amer. Proc. 26:556-558.
MITCHELL, R., and E. HURWITZ. 1965. Suppression of *Pythium debaryanum* by lytic rhizosphere bacteria. Phytopathology 55:156-158.
MIXON, A. C., and E. A. CURL. 1967. Influence of plant residues on *Sclerotium rolfsii* and inhibitory soil microorganisms. Crop Sci. 7:641-644.
MONTEGUT, J. 1960. Value of the dilution method. *In* D. Parkinson and J. S. Waid (eds.), The ecology of soil fungi. Liverpool Univ. Press, Liverpool.
MOREAU, R., and A. VAN EFFENTERRE. 1961. La technique des repliques et l'observation directe des elements fongiques de la rhizosphère. Ann. Inst. Pasteur 101:619-625.
MORRISON, R. H., and D. W. FRENCH. 1969. Direct isolation of *Cylindrocladium floridanum* from soil. Plant Disease Reptr. 53:367-369.
MORTON, D. J., and W. H. STROUBE. 1955. Antagonistic and stimulatory effects of soil microorganisms upon *Sclerotium rolfsii*. Phytopathology 45:417-420.
MOUBASHER, A. H. 1963. Selective effects of fumigation with carbon disulphide on the soil fungus flora. Trans. Brit. Mycol. Soc. 46:338-344.
MUELLER, K. E., and L. W. DURRELL. 1957. Sampling tubes for soil fungi. Phytopathology 47:243.
MUNNECKE, D. E. 1957. Chemical treatment of nursery soils. *In* K. F. Baker (ed.) The U. C. System for producing healthy container-grown plants. Calif. Agr. Exp. Sta. Manual 23.
NADAKAVUKAREN, M. J., and C. E. HORNER. 1959. An alcohol agar medium selective for determining *Verticillium* microsclerotia in soil. Phytopathology 49:527-528.
NASH, S. M., T. CHRISTOU, and W. C. SNYDER. 1961. Existence of *Fusarium solani* f. *phaseoli* as chlamydospores in soil. Phytopathology 51:308-312.
NASH, S. M., and W. C. SNYDER. 1962. Quantitative estimations by plate counts of propagules of the bean root rot *Fusarium* in field soils. Phytopathology 52:567-572.
NELSON, G. A., and O. A. OLSEN. 1964. Methods for estimating numbers of resting sporangia of *Synchytrium endobioticum* in soil. Phytopathology 54:185-186.
NELSON, G. A., and G. SEMENIUK 1963. Persistence of *Corynebacterium insidiosum* in soil. Phytopathology 53:1167-1169.
NELSON, N. 1944. A photometric adaptation of the Somogyi method for the determination of glucose. J. Biol. Chem. 153:375-380.
NEWCOMBE, M. 1960. Some effects of water and anaerobic conditions on *Fusarium oxysporum* f. *cubense* in soil. Trans. Brit. Mycol. Soc. 43:51-59.

NICHOLAS, D. P., D. PARKINSON, and N. A. BURGES. 1965. Studies of fungi in a podzol. II. Application of the soil-sectioning technique to the study of amounts of fungal mycelium in the soil. J. Soil Sci. 16:258-269.

NORTON, D. C. 1953. Linear growth of *Sclerotium bataticola* through soil. Phytopathology 43:633-636.

NOVOGRUDSKY, D. et al. 1937. The influence of bacterization of flaxseed on the susceptibility of seedlings to infection with parasitic fungi. C. R. Acad. Sci. (U.S.S.R.) 14:385-388.

OCANA, G., and P. H. TSAO. 1965. Origin of colonies of *Phytophthora parasitica* in selective pimaricin media in soil dilution plates. Phytopathology 55:1070.

OCANA, G., and P. H. TSAO. 1966. A selective agar medium for the direct isolation and enumeration of *Phytophthora* in soil. Phytopathology 56:893.

OHMS, R. E. 1957. A flotation method for collecting spores of a phycomycetous mycorrhizal parasite from soil. Phytopathology 47:751-752.

OKAFOR, N. 1966. Ecology of micro-organisms on chitin buried in soil. J. Gen. Microbiol. 44(3):311-327.

OLD, K. M., and T. H. NICOLSON. 1962. Use of nylon mesh in studies of soil fungi. Plant Disease Reptr. 46:616.

OLSEN, C. M., and K. F. BAKER. 1968. Selective heat treatment of soil, and its effect on the inhibition of *Rhizoctonia solani* by *Bacillus subtilis*. Phytopathology 58:79-87.

PADY, S. M., C. L. KRAMER, and V. K. PATHAK. 1960. Suppression of fungi by light on media containing rose bengal. Mycologia 52:347-350.

PAHARIA, K. D., and T. KOMMEDAHL. 1956. The effect of time of adding suspensions in soil mycoflora assays. Plant Disease Reptr. 40:1029-1031.

PAPAVIZAS, G. C. 1964. New medium for the isolation of *Thielaviopsis basicola* on dilution plates from soil and rhizosphere. Phytopathology 54:1475-1481.

PAPAVIZAS, G. C. 1967a. Evaluation of various media and antimicrobial agents for isolation of *Fusarium* from soil. Phytopathology 57:848-852.

PAPAVIZAS, G. C. 1967b. Survival of root-infecting fungi in soil. I. A quantitative propagule assay method of observation. Phytopathology 57:1242-1246.

PAPAVIZAS, G. C., and C. B. DAVEY. 1959a. Evaluation of various media and antimicrobial agents for isolation of soil fungi. Soil Sci. 88:112-117.

PAPAVIZAS, G. C., and C. B. DAVEY. 1959b. Isolation of *Rhizoctonia solani* Kuehn from naturally infested and artificially inoculated soils. Plant Disease Reptr. 43:404-410.

PAPAVIZAS, G. C., and C. B. DAVEY. 1960. *Rhizoctonia* disease of bean as affected by decomposing green plant materials and associated microfloras. Phytopathology 50:516-522.

PAPAVIZAS, G. C., and C. B. DAVEY. 1961a. Saprophytic behavior of *Rhizoctonia* in soil. Phytopathology 51:693-699.
PAPAVIZAS, G. C., and C. B. DAVEY. 1961b. Extent and nature of the rhizosphere of *Lupinus*. Plant and Soil 14:215-236.
PAPAVIZAS, G. C., and C. B. DAVEY. 1962. Activity of *Rhizoctonia* in soil as affected by carbon dioxide. Phytopathology 52:759-766.
PAPAVIZAS, G. C., and C. B. DAVEY. 1962. Isolation and pathogenicity of *Rhizoctonia* saprophytically existing in soil. Phytopathology 52:834-840.
PARKER, F. W. 1921. Methods of studying the concentration and composition of soil solution. Soil Sci. 12:209-232.
PARKINSON, D. 1957. New methods for the qualitative and quantitative study of fungi in the rhizosphere. Pedologie 7:(no. sp.) 146-154.
PARKINSON, D., and J. H. CLARKE. 1961. Fungi associated with the seedling roots of *Allium porrum* L. Plant and Soil 13:384-390.
PARKINSON, D., and A. THOMAS. 1965. A comparison of methods for the isolation of fungi from rhizospheres. Can. J. Microbiol. 11:1001-1007.
PARMETER, J. R. JR., and J. R. HOOD. 1961. The use of Fusarium culture filtrate media in the isolation of Fusaria from soil. Phytopathology 51:164-168.
PATEL, M. K. 1926. An improved method of isolating *Pseudomonas tumefaciens* Sm. and Town. Phytopathology 16:577.
PATRICK, Z. A. 1954. The antibiotic activity of soil microorganisms as related to bacterial plant pathogens. Can. J. Bot. 32:705-735.
PETERSON, E. A. 1954. A study of cross antagonisms among some actinomycetes active against *Streptomyces scabies* and *Helminthosporium sativum*. Antib. and Chemo. 4:145-149.
PISANO, M. A., J. M. CARSON, and A. I. FLEISCHMAN. 1962. A wheat germ agar for the cultivation of fungi. Can. J. Microbiol. 8:277-279.
PORTER, J. N., J. J. WILHELM, and H. D. TRESNER. 1960. Method for the preferential isolation of actinomycetes from soil. Appl. Microbiol. 8:174-178.
PORTER, L. K. 1965. Enzymes, p. 1536-1549. *In* Methods of soil analysis, Part II. Amer. Soc. Agron., Inc. Madison, Wis.
PRATT, R., and J. DUFRENOY. 1953. Antibiotics. J. B. Lippincott, Co., Philadelphia, Penn.
PRIDHAM, T. G. et al. 1956. Antibiotics against plant disease. I. Laboratory and greenhouse survey. Phytopathology 46:568-575.
RANGASWAMI, G., and S. ETHIRAJ. 1962. Antibiotic production by *Streptomyces* sp. in unamended soil. Phytopathology 52:989-992.
RANNEY, C. D. 1956. Design and construction of a compact battery of constant temperature tanks for cotton seedling disease investigations. Plant Disease Reptr. 40:559-563.

RAPER, K. B., and C. THOM. 1949. Manual of the penicillia. The Williams and Wilkins Co., Baltimore, Md.
RAUTENSHTEYN, V. I., and N. D. KOFANOVA. 1957. On the isolation of actinophages from soil. Mikrobiologiya (A.I.B.S. trans.) 26:318-326.
REEVE, R. C., and E. J. DOERING. 1965. Sampling the soil solution for salinity appraisal. Soil Sci. 99:339-344.
REHACEK, Z. 1959. Isolation of actinomycetes and determination of the number of their spores in soil. Microbiologiya (A.I.B.S. trans.) 28:220-225.
REYES, A. A., and J. E. MITCHELL. 1962. Growth response of several isolates of *Fusarium* in rhizospheres of host and nonhost plants. Phytopathology 52:1196-1200.
RHOADES, H. L., and M. B. LINDFORD. 1959. Control of Pythium root rot by the nematode *Aphelenchus avenae*. Plant Disease Reptr. 43:323-328.
RICHARDS, L. A. 1941. A pressure-membrane extraction apparatus for soil solution. Soil Sci. 51:377-386.
RICHARDS, L. A. 1947. Pressure-membrane apparatus-construction and use. Agr. Engin. 28:451-454.
RICHARDS, L. A. 1949. Filter funnels for soil extracts. Agron. J. 41:446.
RIKER, A. J., and R. S. RIKER. 1936. Introduction to research on plant diseases. John S. Swift Co., St. Louis, Mo.
RISBETH, J. 1950. Observations on the biology of *Fomes annosus*, with particular reference to East Anglian pine plantations. I. The outbreaks of disease and ecological status of the fungus. Ann. Bot. 14:365-383.
ROBERTSON, N. F. 1954. Studies on the mycorrhiza of *Pinus sylvestris*. I. New Phytol. 53:253-283.
ROBINSON, J. B., and C. T. CORKE. 1959. Preliminary studies on the distribution of actinophages in soil. Can. J. Microbiol. 5:479-484.
RODRIGUEZ-KABANA, R. 1967. An improved method for assessing soil fungus population density. Plant and Soil 26:393-396.
RODRIGUEZ-KABANA, R. 1969. Enzymatic interactions of *Sclerotium rolfsii* and *Trichoderma viride* in mixed soil culture. Phytopathology 59:910-921.
RODRIGUEZ-KABANA, R., and E. A. CURL. 1968. Saccharase activity of *Sclerotium rolfsii* in soil and the mechanism of antagonistic action by *Trichoderma viride*. Phytopathology 58:985-992.
RODRIGUEZ-KABANA, R., E. A. CURL, and H. H. FUNDERBURK, JR. 1967. Effect of paraquat on growth of *Sclerotium rolfsii* in liquid culture and soil. Phytopathology 57:911-915.
ROGERS, C. H. 1936. Apparatus and procedure for separating cotton root-rot sclerotia from soil samples. J. Agr. Res. 52:73-79.
ROHDE, P. A., ed. 1968. BBL manual of products and laboratory procedures. 5th Ed., Bioquest, Div. of Becton, Dickinson and Co., Cockeysville, Md.
ROHDE, R. A., and W. R. JENKINS. 1958. The chemical basis of resistance of

asparagus to the nematode *Trichodorus christiei.* Phytopathology 48:463.
ROSE, R. E., and J. G. MILLER. 1954. Some sampling variations in soil fungal numbers. J. Gen. Microbiol. 10:1-10.
ROSSI, G., and S. RICCARDO. 1927. L'seame microscopico e bacteriologico diretto del terreno agrario. Nuovi Ann. Minist. Agric. 7:457-470.
ROUATT, J. W., and R. G. ATKINSON. 1950. The effect of the incorporation of certain cover crops on the microbiological balance of potato scab infested soil. Can. J. Res. 28(c):140-152.
ROVIRA, A. D. 1956. Plant root excretions in relation to the rhizosphere effect. I, II, and III. Plant and Soil 7:178-217.
ROVIRA, A. D. 1959. Root excretions in relation to the rhizosphere effect. IV. Influence of plant species, age of plant, light, temperature, and calcium nutrition on exudation. Plant and Soil 11:53-64.
ROVIRA, A. D., and J. R. HARRIS. 1961. Plant root excretions in relation to the rhizosphere effect. V. The exudation of B-group vitamins. Plant and Soil 14:199-214.
ROZHDESTVENSKII, V. A. 1961. Primenenie spetsial'nogo shchupa dlya otbora prob pri mikrobioligicheskikh issledovaniyakh pochv. Lab. Delo 9:52-53. (Referat Zhur., Biol. 1962. No. 3B361).
RUSSELL, P. 1956. A selective medium for the isolation of basidiomycetes. Nature 177:1038-1039.
SCHENCK, N. C., and E. A. CURL. 1961. A short quantitative method for estimating the population density of soil fungi. Proc., Soil and Crop Sci. Soc. Florida 21:13-17.
SCHMIDT, E. L., and R. O. BANKOLE. 1962. Detection of *Aspergillus flavus* in soil by immunofluorescent staining. Science 136:776-777.
SCHMITTHENNER, A. F. 1962. Isolation of *Pythium* from soil particles. Phytopathology 52:1133-1138.
SCHMITTHENNER, A. F., and L. E. WILLIAMS. 1958. Methods for analysis of soil-borne plant pathogens and associated soil fungi. Ohio Agr. Exp. Sta., Botany and Plant Pathology Mimeo. Series No. 29.
SCHREIBER, L. R., and R. J. GREEN, JR. 1962. Comparative survival of mycelium, conidia, and microsclerotia of *Verticillium albo-atrum* in mineral soil. Phytopathology 52:288-289.
SCHROTH, M. N., and W. C. SNYDER. 1961. Effect of host exudates on chlamydospore germination of the bean root rot fungus, *Fusarium solani* f. *phaseoli.* Phytopathology 51:389-393.
SCHROTH, M. N., J. P. THOMPSON, and D. C. HILDEBRAND. 1965. Isolation of *Agrobacterium tumefaciens - A. radiobacter* Group from soil. Phytopathology 55:645-647.
SCOTT, W. W., *ed.* 1925. Standard methods of chemical analysis. D. Van Nostrand Co., Inc. New York, N. Y.

SEWELL, G. W. F. 1959. Studies of fungi in a Calluna-heathland soil. II. By the complementary use of several isolation methods. Trans. Brit. Mycol. Soc. 42:354-369.

SHAPIRO, R. E., G. S. TAYLOR, and G. W. VOLK. 1956. Soil oxygen contents and ion uptake by corn. Soil Sci. Soc. Am., Proc. 20:192-197.

SHARVELLE, E. G. 1961. The nature and uses of modern fungicides. Burgess Publishing Co., Minneapolis, Minn.

SHEARD, R. W. 1961. A note on an aid for rapid vacuum filtration. Can. J. Soil Sci. 41:260.

SHERWOOD, R. T. 1958. Aphanomyces root rot of garden pea. PhD. Thesis. Univ. of Wis.

SHIVE, J. W., and W. R. ROBBINS. 1942. Methods of growing plants in solution and sand cultures. New Jersey Agr. Exp. Sta. Bul. 636.

SINGH, K. G. 1965. Comparison of techniques for the isolation of root-infecting fungi. Nature 206:1169-1170.

SINGH, R. S., and J. E. MITCHELL. 1961. A selective method for isolation and measuring the population of *Pythium* in soil. Phytopathology 51:440-444.

SKINNER, F. A., P. C. T. JONES, and J. E. MOLLISON. 1952. A comparison of a direct- and plate-counting technique for the quantitative estimation of soil microorganisms. J. Gen. Microbiol. 6:261-271.

SKUJINS, J. J. 1967. Enzymes in soil, p. 317-414. *In* A. D. McLaren and G. H. Peterson (eds.), Marcel Dekker, Inc., New York.

SKUJINS, J. J., and A. D. MCLAREN. 1968. Persistence of enzymatic activities in stored and geologically preserved soils. Enzymologia 34:213-225.

SLAGG, C. M., and H. FELLOWS. 1947. Effects of certain soil fungi and their by-products on *Ophiobolus graminis*. J. Agr. Res. 75:279-293.

SLEETH, B. 1945. Agar medium and technique for isolating *Pythium* free of bacteria. Phytopathology 35:1030-1031.

SMITH, J. G. 1960. The influence of antagonistic fungi on *Thielaviopsis basicola* (Berk. et Br.) Ferraris. Acta Botanica Neerlandica 9:59-118.

SNYDER, W. C., and H. N. HANSEN. 1947. Advantage of natural media and environments in the culture of fungi. Phytopathology 37:420-421.

SOMOGYI, M. 1945. A new reagent for the determination of sugars. J. Biol. Chem. 160:61-68.

SOULIDES, D. A. 1964. Antibiotics in soils: VI. Determination of microquantities of antibiotics in soil. Soil Sci. 97:286-289.

STANSLY, P. G. 1947. A bacterial spray apparatus useful in searching for antibiotic-producing microorganisms. J. Bact. 54:443-445.

STARKEY, R. L. 1938. Some influences of the development of higher plants upon the microorganisms in the soil. VI. Microscopic examination of the rhizosphere. Soil Sci. 45:207-249.

STATEN, G., and J. F. COLE, JR. 1948. The effect of preplanting irrigation on pathogenicity of *Rhizoctonia solani* in seedling cotton. Phytopathology 38:661-664.
STEELE, A. E. 1967. A constant temperature bath for pot-grown plants. Plant Disease Reptr. 51:171-173.
STEINER, G. W., and R. D. WATSON. 1965. Use of surfactants in the soil dilution and plant count method. Phytopathology 55:728-730.
STESSEL, G. J., C. LEBEN, and G. W. KEITT. 1953. Screening tests designed to discover antibiotics suitable for plant disease control. Mycologia 45:325-334.
STEVENSON, I. L. 1956. Antibiotic activity of actinomycetes in soil and their controlling effects on root-rot of wheat. J. Gen. Microbiol. 14:440-448.
STEVENSON, I. L. 1956. Antibiotic activity of actinomycetes in soil as demonstrated by direct observation techniques. J. Gen. Microbiol. 15:372-380.
STEVENSON, I. L. 1958. The effect of sonic vibration on the bacterial plate count of soil. Plant and Soil 10:1-8.
STEVENSON, I. L., and A. G. LOCHHEAD. 1953. The use of a percolation technique in studying antibiotic production in soil. Can. J. Bot. 31:23-27.
STONER, M. F., and R. J. COOK. 1967. Use of phytoactin in population studies of *Fusarium roseum "Culmorum"* in field soils. Phytopathology 57:102.
STOTZKY, G. 1965. Microbial respiration, p. 1550-1572. *In* Methods of soil analysis, Part II. American Soc. Agron., Inc., Madison, Wis.
STOTZKY, G., R. D. GOOS, and M. I. TIMONIN. 1962. Microbial changes occurring in soil as a result of storage. Plant and Soil 16:1-18.
STOTZKY, G., T. M. RYAN, and J. L. MORTENSEN. 1958. Apparatus for studying biochemical transformation in incubated soils. Soil Sci. Soc. Am., Proc. 22:270-271.
STOVER, R. H. 1958a. Studies on *Fusarium* wilt of bananas. II. Some factors influencing survival and saprophytic multiplication of *F. oxysporum* f. *cubense* in soil. Can. J. Bot. 36:311-324.
STOVER, R. H. 1958b. Studies on *Fusarium* wilt of bananas. III. Influence of soil fungitoxins on behavior of *F. oxysporum* f. *cubense* in soil extracts and diffusates. Can. J. Bot. 36:439-453.
STOVER, R. H., and B. H. WAITE. 1953. An improved method of isolating *Fusarium* spp. from plant tissue. Phytopathology 43:700-701.
STOVER, R. H., and B. H. WAITE. 1954. Colonization of banana roots by *Fusarium oxysporum* f. *cubense* and other soil fungi. Phytopathology 44:689-693.
STREETS, R. B. 1937. Phymatotrichum (cotton or Texas) root rot in Arizona. Ariz. Agr. Exp. Sta. Tech. Bul. 71.

STRUGGER, S. 1948. Fluorescence microscope examination of bacteria in soil. Can. J. Res. 26(c):188-193.
SZYBALSKI, W. 1952. Gradient plate technique for study of bacterial resistance. Science 116:46-48.
TCHAN, Y. T. 1952. Counting soil algae by direct fluorescence microscope. Nature 170:328-329.
TEAKLE, D. S., and C. E. YARWOOD. 1962. Improved recovery of tobacco necrosis virus from roots by means of *Olpidium brassicae.* Phytopathology 52:366.
TELIZ-ORTIZ, M., and W. H. BURKHOLDER. 1960. A strain of *Pseudomonas fluorescens* antagonistic to *Pseudomonas phaseolicola* and other bacterial plant pathogens. Phytopathology 50:119-123.
THIES, W. G., and R. F. PATTON. 1970. An evaluation of propagules of *Cylindrocladium scoparium* in soil by direct isolation. Phytopathology 60:599-601.
THOMAS, A., D. P. NICHOLAS, and D. PARKINSON. 1965. Modifications of the agar film technique for assaying lengths of mycelium in soil. Nature 205:105.
THORNTON, H. G. 1922. On the development of a standardized agar medium for counting soil bacteria, with especial regard to the repression of spreading colonies. Ann. Appl. Biol. 9:241-274.
THORNTON, R. H. 1952. The screened immersion plate: A method of isolating soil micro-organisms. Research 5:190-191.
TIMONIN, M. I. 1940. The interaction of higher plants and soil microorganisms. I. Microbial population of rhizosphere of seedlings of certain cultivated plants. Can. J. Res. 18(c):307-317.
TIMONIN, M. I. 1941. The interactions of higher plants and soil microorganisms. III. Effect of by-products of plant growth on activity of fungi and actinomycetes. Soil Sci. 52:395-413.
TISDALE, S. L., and W. L. NELSON. 1966. Soil fertility and fertilizers. 2nd Ed., The Macmillan Co., New York. N. Y.
TOLMSOFF, W. J. 1959. The isolation of nematode trapping fungi from Oregon soils. Phytopathology 49:113-114.
TREADWELL, F. P.1937. Analytical chemistry. John Wiley and Sons, Inc., New York, N. Y.
TRIBE, H. T. 1957. On the parasitism of *Sclerotinia trifoliorum* by *Coniothyrium minitans.* Trans. Brit. Mycol. Soc. 40:489-499.
TRIBE, H. T. 1960. Decomposition of buried cellulose film with special reference to the ecology of certain soil fungi. *In* D. H. Parkinson and J. S. Waid (ed.), The ecology of soil fungi, Liverpool Univ. Press.
TSAO, P. H. 1960. A serial dilution end-point method for estimating disease potentials of citrus Phytophthoras in soil. Phytopathology 50:717-724.

TSAO, P. H. 1964. Effect of certain fungal isolation agar media on *Thielaviopsis basicola* and on its recovery in soil dilution plates. Phytopathology 54:548-555.
TSAO, P. H., and S. D. VAN GUNDY. 1960. Pathogenicity on citrus of *Thielaviopsis basicola* and its isolation from field roots. Phytopathology 50:86.
TSAO, P. H., C. LEBEN, and G. W. KEITT. 1960. An enrichment method for isolating actinomycetes that produce diffusible antifungal antibiotics. Phytopathology 50:88-89.
TUITE, J. 1969. Plant pathological methods, fungi and bacteria. Burgess Publ. Co., Minneapolis, Minn.
TVEIT, M., and M. B. MOORE. 1954. Isolates of *Chaetomium* that protect oats from *Helminthosporium victoriae.* Phytopathology 44:686-689.
TVEIT, M., and R. K. S. WOOD. 1955. The control of Fusarium blight in oat seedlings with antagonistic species of *Chaetomium.* Ann. Appl. Biol. 43:538-552.
UMBREIT, W. W., R. H. BURRIS, and J. F. STAUFFER. 1964. Manometric techniques. Burgess Publishing Co., Minneapolis, Minn.
UNGER, H. 1957. Ein Bohrer zur Entnahme von Bodenproben für mikrobiologische Untersuchungen. Arch. Mikrobiol, Bd. 27:429-432.
VAN BAVEL, C. H. M. 1965. Composition of soil atmosphere. *In* Methods of soil analysis, Part I. Amer. Soc. Agron., Inc., Madison, Wis.
VASUDEVA, R. S., G. P. SINGH, and M. R. S. IYENGAR. 1962. Biological activity of bulbiformin in soil. Ann. Appl. Biol. 50:113-119.
VASUDEVA, R. S., T. V. SUBBAIAH, M. L. N. SASTRY, G. RANGASWAMY, and M. R. S. IYENGAR. 1958. "Bulbiformin", an antibiotic produced by *Bacillus subtilis.* Ann. Appl. Biol. 46:336-345.
VUJICIC, R., and D. PARK. 1964. Behavior of *Phytophthora erythroseptica* in soil. Trans. Brit. Mycol. Soc. 47:455-458.
WAID, J. S. 1956. Root dissection: A method of studying the distribution of active mycelia within root tissue. Nature 178:1477-1478.
WAID, J. S. 1957. Distribution of fungi within the decomposing tissues of rye grass roots. Trans. Brit. Mycol. Soc. 40:391-406.
WAID, J. S., and M. J. WOODMAN. 1957. A method of estimating hyphal activity in soil. Pedologie 7:155-158.
WAKSMAN, S. A. 1916. Do fungi live and produce mycelium in the soil? Sci., N.S. 44:320-322.
WAKSMAN, S. A. 1945. Microbial antagonisms and antibiotic substances. The Commonwealth Fund, New York, N. Y.
WAKSMAN, S. A. 1950. The actinomycetes—their nature, occurrence, activities and importance. Chronica Botanica Co., Waltham, Mass.
WAKSMAN, S. A., and E. B. FRED. 1922. A tentative outline of the plate

method for determining the number of microorganisms in the soil. Soil Sci. 14:27-28.
WAKSMAN, S. A., and A. SCHATZ. 1946. Soil enrichment and development of antagonistic microorganisms. J. Bact. 51:305-316.
WARCUP, J. H. 1950. The soil-plate method for isolation of fungi from soil. Nature 166:117-118.
WARCUP, J. H. 1951. Soil-steaming: a selective method for the isolation of Ascomycetes from soil. Trans. Brit. Mycol. Soc. 34:515-518.
WARCUP, J. H. 1955. Isolation of fungi from hyphae present in soil. Nature 175:953-954.
WARCUP, J. H. 1955. On the origin of colonies of fungi developing on soil dilution plates. Trans. Brit. Mycol. Soc. 38:298-301.
WARCUP, J. H. 1959. Studies on basidiomycetes in soil. Trans. Brit. Mycol. Soc. 42:45-52.
WARCUP, J. H., and K. F. BAKER. 1963. Occurrence of dormant ascospores in soil. Nature 197:1317-1318.
WARCUP, J. H., and P. H. B. TALBOT. 1962. Ecology and identity of mycelia isolated from soil. Trans. Brit. Mycol. Soc. 45:495-518.
WARREN, H. B., JR., J. F. PROKOP, and W. E. GRUNDY. 1955. Nonsynthetic media for antibiotic producing actinomycetes. Antib. and Chemoth. 5:6-12.
WATSON, R. D. 1960. Soil washing improves the value of the soil dilution and the plate count method of estimating populations of soil fungi. Phytopathology 50:792-794.
WEBER, D. F., J. D. MENZIES, and J. W. PADEN. 1963. Streptomycin resistance—a technique for the direct assay of potato scab organisms from soil. Phytopathology 53:1258-1260.
WEINBERG, E. D. 1957. Double-gradient agar plates. Science 125:196.
WEINHOLD, A. R., and T. BOWMAN. 1965. Influence of substrate on activity of a bacterium antagonistic to *Streptomyces scabies.* Phytopathology 55:126.
WEINDLING, R. 1932. *Trichoderma lignorum* as a parasite of other soil fungi. Phytopathology 22:837-845.
WEINDLING, R., and H. S. FAWCETT. 1936. Experiments in the control of Rhizoctonia damping-off of citrus seedlings. Hilgardia 10:1-16.
WILHELM, S. 1950. Vertical distribution of *Verticillium albo-atrum* in soils. Phytopathology 40:368-375.
WILHELM, S. 1956. A sand-culture technique for the isolation of fungi associated with roots. Phytopathology 46:293-295.
WILLIAMS, G. H., and J. H. WESTERN. 1965. The biology of *Sclerotinia trifoliorum* Eriks, and other species of sclerotium-forming fungi. I. Apothecium formation from sclerotia. Ann. Appl. Biol. 56:253-260.

WILLIAMS, L. E. 1967. Origin of fungal colonies on soil dilution plates. Phytopathology 57:836.
WILLIAMS, L. E., and A. F. SCHMITTHENNER. 1962. Effect of crop rotation on soil fungus populations. Phytopathology 52:241-247.
WILLIAMS, L. E., and G. M. WILLIS. 1962. Agar-ring method for in vitro studies of fungistatic activity. Phytopathology 52:368-369.
WILLIAMS, S. T., and F. L. DAVIES. 1965. Use of antibiotics for selective isolation and enumeration of actinomycetes in soil. J. Gen. Microbiol. 38:251-261.
WILLIAMS, S. T., D. PARKINSON, and N. A. BURGES. 1965. An examination of the soil washing technique by its application to several soils. Plant and Soil 22:167-186.
WILLOUGHBY, L. G. 1956. Studies on soil chytrids. I. *Rhizidium richmondense* sp. nov. and its parasites. Trans. Brit. Mycol. Soc. 39:125-141.
WILLSON, D. and H. S. FOREST. 1957. An exploratory study of soil algae. Ecology 38:309-313.
WITKAMP, M., and R. L. STARKEY. 1956. Tests of some methods for detecting antibiotics in soil. Soil Sci. Soc. Amer., Proc. 20:500-504.
WOOD, F. A., and R. D. WILCOXSON. 1960. Another screened immersion plate for isolating fungi from soil. Plant Disease Reptr. 44:594.
WOOD, R. K. S. 1951. The control of diseases of lettuce by the use of antagonistic organisms. I. The control of *Botrytis cinerea* Pers. Ann. Appl. Biol. 38:203-216.
WORF, G. L., and D. J. HAGEDORN. 1961. A technique for studying relative soil populations of two *Fusarium* pathogens of garden peas. Phytopathology 51:805-806.
WRIGHT, E. 1945. Relation of macrofungi and microorganisms of soils to damping-off of broadleaf seedlings. J. Agr. Res. 70:133-141.
WRIGHT, J. M. 1952. Production of gliotoxin in unsterilized soil. Nature 170:673-674.
WRIGHT, J. M. 1956. Biological control of a soil-borne Pythium infection by seed inoculation. Plant and Soil 8:132-140.
YAMAGUCHI, M., F. D. HOWARD, D. L. HUGHES, and W. J. FLOCKER. 1962. An improved technique for sampling and analysis of soil atmospheres. Soil Sci. Soc. Am., Proc. 26:512-513.
YARWOOD, C. E. 1946. Isolation of *Thielaviopsis basicola* from soil by means of carrot disks. Mycologia 38:346-348.
YARWOOD, C. E. 1960. Release and preservation of virus by roots. Phytopathology 50:111-114.
ZAK, B. 1962. Direct isolation of fungal symbionts from pine mycorrhizae. Phytopathology 52:34.

ZAN, K. 1962. Activity of *Phytophthora infestans* in soil in relation to tuber infection. Trans. Brit. Mycol. Soc. 45:205-221.

ZENTMYER, G. A., J. D. GILPATRICK, and W. A. THORN. 1960. Methods of isolating *Phytophthora cinnamomi* from soil and from host tissue. Phytopathology 50:87.

ZENTMYER, G. A., and C. R. THOMPSON. 1967. The effect of saponins from alfalfa on *Phytophthora cinnamomi* in relation to control of root rot of avocado. Phytopathology 57:1278-1279.

ZVYAGINTSEV, D. G. 1959. Adsorption of microorganisms by glass surfaces. Mikrobiologiya (A.I.B.S. trans.) 28:104-108.

ZVYAGINTSEV, D. G. 1964. The use of quenchers in the investigation of soil microorganisms by fluorescence microscopy. Mikrobiologiya (A.I.B.S. trans.) 32:622-625.

SUBJECT INDEX

Acrasiales, 27
Acrostalagmus roseus, 147
Actinomycetes, enumeration and observation *in situ*, 36-47; identification, 12; isolation from soil, 12-14; media for isolation, 191-195
Actinophage, 150-151
Agar media, see culture media
Agrobacterium rhizogenes, 59, 75; *A. tumefaciens*, 59, 74, 75, 155, 181, 199, 200
Algae, 197; enumeration and observation *in situ*, 42; isolation from soil, 33-35
Alternaria sp., 43; *A. solani*, 155; *A. tenuis*, 166, 170
Anderson air sampler, 71
Antibiotic activity, 153-166
Antibiotics, activity in soil, 167-175; assay, 157-166; media for production, 203-206; production in soil, 169-175; recovery from soil, 168-171; stability in soil, 167-169; treatment of seed, 180-181
Antagonism, criteria, 153-154; methods for determining, 154-157
Antagonistic microorganisms, added to soil, 177-179; isolation, 139-152; stimulation by chemicals, 185-186; testing for antagonism, 153-166; treatment of seed, 179-180

Aphanomyces cochlioides, 181, 183; *A. euteiches*, 75
Aphelenchus avenae, 145, 177
Armillaria mellea, 145
Arthrobacter sp., 180
Ascomycetes, 28
Aspergillus sp., 25, 146; *A. flavus*, 179; *A. fumigatus*, 183; *A. niger*, 181

Bacillus cereus, 13; *B. mesentericus*, 13; *B. mycoides*, 13; *B. subtilis*, 13, 169, 204
Bacteria, autochthonous, 39; enumeration and observation *in situ*, 36-47; identification, 12; isolation from soil, 7-11; media for cultivation, 187-191; media for isolation, 191-193; vital staining, 39
Bacteriophage, 148-150
Basidiomycetes, 202; isolation from soil, 29; isolation from woody tissue, 78
Biological control, 176-186
Botrytis cinerea, 147, 155
Buried-slide technique, 24, 39-41, 55, 174-175

Cambridge method, 125
Carbon dioxide, collection apparatus, 104-107; effect on fungi, 87-88; measurement of concentration in soil, 85-88; 103-107
Cellulose decomposing fungi, 198; isolation from soil, 32-33

237

Chaetomium sp., 179, 180
Chalaropsis thielavioides, 70
Chromatographic techniques, 86-87; 135, 171
Chytrids, 27
Cladosporium sp., 78
Claviceps purpurea, 147
Colletotrichum sp., 180
Competitive Saprophytic Ability (CSA), 124-125
Coniothyrium minitans, 146, 147
Corynebacterium insidiosum, 119; *C. michiganese,* 181
Crop residue amendments, 180-183
Crop rotation, 183-184
Culture filtrates, added to soil, 180-181; assay, 157-166; treatment of seed, 180-181
Culture media, 187-208; actinomycete-fermentation medium, 204; alcohol agar, 71; arginine-glycerol-salt agar, 193; bacitracin-fermentation medium, 205; Benedict agar, 193; Bristol's sodium nitrate solution, modified, 197; cellulose agar, 197; chitin agar, 194; chloromycetin-fermentation medium, 205; claviformin-fermentation medium, 205; corn meal agar, 187; Crone's solution, 206; Czapek's sucrose-nitrate agar, 187; dextrose agar, 188; dextrose-peptone-yeast extract agar, 195; egg albumin agar, 194; endomycin-fermentation medium, 205; gallic acid agar, 198; gliotoxin-fermentation medium, 205; glucose-asparagine agar, 194; glycerol-asparaginate agar, 194; heart infusion agar, 188; Hoagland's solution I, 206; Hoagland's solution II, 207; Hutchinson's agar, modified, 191; Jensen's agar, 194; Kerr's agar, modified, 198; lima bean agar, 188; maize meal agar, 199; malt extract agar, 188; mannitol-nitrate agar, 199; milk agar, 30; natural media, 188; nutrient agar, 189; oatmeal agar, 189; Ohio agar, 196; Patel's agar, 199; PCNB agar, 200; PCNB agar, modified, 200; peptone-dextrose-phosfon agar, 197; peptone-dextrose-rose bengal agar, 196, 197; potato-dextorse agar, 189, 197; potato plugs, 190; Reischer's agar, 201; Richard's agar, 201; rose bengal agar, modified, 202; Russell's agar, 202; Shive and Robbin's solution I, 207; Shive and Robbin's solution II, 207; Sleeth's agar, 76; sodium albuminate agar, 192; soil extract agar, 191, 192; soil extract agar, dilute, 202; soybean meal-glucose agar, 195; starch-casein agar, 195; test medium, 206; Thornton's standardized agar, 193; trypticase-soy agar, 190; tryptose agar, 190; tyrosine-casein-nitrate agar, 203; V-8 juice agar, 190; VDYA-PCNB agar, 203; wheat germ-glucose agar, 191
Culture solutions for higher plants, 206-208
Curvularia ramosa, 125
Cylindrocladium scoparium, 60

Dactyella spermatophaga, 145
Dehydrogenase activity, 108-110
Dilution-plate technique, 7-9, 12, 14-15, 36, 49-50, 140-144
Disinfectants, 73-74
Direct microscopic examination of soil, 36-39

Enzyme activity, 108-115
Erwinia amylovora, 155; E. caratovora, 155
Escherichia coli, 27

Fomes annosus, 60, 78, 79, 200
Fungi, enumeration and observation in situ, 36-47; isolation from soil, 14-33; media for cultivation, 187-191; medium for isolation, 195-203
Fusarium sp., 2, 25, 37, 51, 52, 58, 60, 61, 62, 74, 75, 122, 127, 136, 180, 201; F. culmorum, 125, 155; F. nivale, 180, 181; F. oxysporum f. conglutinans, 61; F. oxysporum f. cubense, 118, 155, 174; F. oxysporum f. lycopersici, 181; F. oxysporum f. pisi, 119, 120, 155; F. oxysporum f. tabaci, 143; F. roseum, 141, 180; F. roseum f. culmorum, 61; F. roseum f. gibbosum, 61; F. solani, 37, 145; F. solani f. phaseoli, 61, 137, 182; F. solani f. pisi, 119, 120

Geotrichum candidum, 62
Gradient-plate technique, 164-165

Helminthosporium sativum, 37, 43, 62, 74, 125, 155, 156, 180; H. victoriae, 179
Hyphae, enumeration and observation in situ, 36-47; isolation from soil, 20-23, 29-30; measurement of length, 39; types in soil, 47

Immersion tubes, 19-20, 127
Immunoflorescent-staining, 43
Inoculum density, 125-129
Inoculum potential, 125-129

Lignin-decomposing fungi, 33

Mortierella sp., 25

Mucor sp., 25, 62, 76
Mycorrhizae, isolation from roots, 79-80; isolation from soil, 28-30
Nematode-trapping fungi, 145; isolation from soil, 30-32

Olpidium brassicae, 81
Ophiobolus graminis, 74, 125, 155, 177, 179, 184
Organic amendments, 180-183
Oxygen, control, 87-88; effect on fungi, 87-88; measurement in soil, 87-88, 102-103

Panagrellus redivivus, 31
Pedoscopes, 41
Penicillium sp., 25, 78, 146, 180, 183, 205; P. cyclopium, 171; P. purpurogenum, 171; P. vermiculatum, 145, 148, 149
Perfusion apparatus, 93-96, 133-134, 151, 170
Periconia sp., 25
Phymatotrichum omnivorum, 62, 182
Phytophthora sp., 74, 75, 76, 198; P. cinnamomi, 63, 64, 65, 176, 199; P. citrophthora, 63, 76, 126; P. cryptogae, 76; P. erythroseptica, 118; P. fragariae, 76; P. infestans, 65; P. megasperma var. sojae, 63; P. parasitica, 63, 76, 126, 145; P. parasitica var. nicotianae, 64, 87, 88
Plant pathogens, biological control, 176-186; isolation from roots, 73-81; isolation from soil, 58-72; media for cultivation, 187-191; media for isolation, 197-203
Plicaria fulva, 146
Predacious fungi, 30-32, 145
Pressure-membrane apparatus, 99-101

Pseudomonas mors-prunorum, 148, 149, 150; *Ps. syringae,* 156; *Ps. tabaci,* 163

Pythium sp., 25, 58, 64, 65, 66, 67, 76, 77, 122, 136, 145, 154, 158, 180, 198, 199, 201; *P. aphanidermatum,* 67; *P. arrhenomanes,* 145, 154, 155, 177, 178, 184; *P. debaryanum,* 180; *P. graminicola,* 141, 174; *P. ultimum,* 80, 180

Respiration, 102-107

Rhizoctonia sp., 25, 58, 122, 136, 154, 182; *R. solani,* 67, 68, 87, 127, 141, 145, 146, 149, 155, 185, 186, 202

Rhizopus sp., 76, 145, 181

Rhizosphere, 48-57; artificial, 137-138; effect, 48; sampler, 52-54

Rhizoplane, 48-49

Root exudates, 130-138

Root observation boxes, 56-57

Root washing procedure, 51-52

R/S ratio, 49

Saccharase activity, 110-115

Sclerotia, isolation from soil, 68-69; parasitism of, 145-148

Sclerotinia borealis, 147; *S. fructicola,* 165; *S. sclerotiorum,* 146, 147; *S. trifoliorum,* 146, 147

Sclerotium bataticola, 74, 116; *S. cepivorum,* 69; *S. rolfsii,* 68, 69, 112, 113, 114, 115, 145, 146, 147, 155, 182

Seed sterilization, 130-135

Soil atmosphere, analysis, 85-86; control 87-88; sampler, 86

Soil extract, methods for obtaining, 92-101; use in culture media, 10-11

Soil fungistasis, 121-124

Soil microorganisms, enumeration and observation *in situ,* 36-47; enzyme activity, 102-115; growth in soil, 116-129; isolation from rhizosphere, 49-51; isolation from rhizoplane, 51-52; isolation from soil, 6-35; parasitism among, 144-148; predation among, 144-145; screening for antagonism, 139-152; survival in soil, 116-129; testing for antagonism, 153-166; variation and distribution, 1-3

Soil moisture, determination, 82; methods for control, 82-84

Soil pH, 89

Soil-plate method, 15-17, 24, 25, 50-51, 65

Soil-recolonization tube, 116-117

Soil samples, 1-5, 52-55

Soil solution, extraction, 92-101; isolation of antibiotics, 170

Soil sectioning, 44-47

Soil sterilization, 90, 91, 184-185

Soil temperature tanks, 85

Soil-washing apparatus, 22, 23, 51

Split-root technique, 83-84

Sporotrichum sp., 25

Spray apparatus, 142-143

Stains, acridine-orange, 39-40; carbol erythrosin, 37, 118; lacto-fuchsin, 38; lactophenol trypan blue, 123; phenolic aniline blue, 38, 42, 56; phenolic erythrosin, 40; phenolic rose bengal, 40, 55, 120, 123; rose bengal, 37

Sterile plant culture, 130-135

Streptomyces sp., 146, 156, 205; *S. antibioticus,* 152; *S. ipomoeae,* 150, 152; *S. lavendulae,* 152; *S. scabies,* 69, 70, 77, 150, 152, 155, 183, 185, 203; *S. venezuelae,* 205

Suction-plate apparatus, 96-97
Synchytrium endobioticum, 43, 44

Talaromyces sp., 25
Thielaviopsis basicola, 37, 70, 77, 145, 202, 203
Torula sp., 25
Trichocladium sp., 25
Trichoderma sp., 25, 60, 145, 146, 148, 180; *T. lignorum,* 145, 185; *T. viride,* 78, 112, 113, 114, 115, 169, 180, 183, 184, 185, 205
Trichodorus christiei, 176
Trichothecium sp., 43

Trinacrium subtilis, 145

Verticillium albo-atrum, 71, 77, 78, 119, 129, 141, 145, 156, 203; *V. dahliae,* 71, 155
Virus, release from roots, 81; transmission by *Olpidium,* 81
Vital staining, 39, 40

Warburg respirometer, 102-103
Wood rotting fungi, 78-79

Xanthomonas phaseoli, 149, 150; *X. translucens,* 181

Yeasts, 28-29

SENIOR AUTHOR INDEX

Aberdeen, J. E. C., 16, 17
Adams, M. H., 150
Adams, P. B., 123
Agnihothrudu, V., 24
Alexopoulos, C. J., 155
Allen, O. N., 11, 33, 93, 191, 197
Allison, L. E., 91
Andal, R., 135
Anderson, A. L., 18
Anderson, E. J., 65
Anker, H. S., 204
Anwar, A. A., 155
Aristovskaya, T. V., 41
Ark, P. A., 59, 74
Ashworth, L. J. Jr., 186
Audus, L. F., 93, 94, 133, 170
Ayers, W. A., 131, 132, 133, 135, 136

Baker, K. F., 90
Baron, A. L., 204
Bartha, R., 106, 107
Bhat, J. V., 9, 11, 191
Blair, I. D., 117
Bliss, D. E., 184
Bloom, J. R., 89
Boosalis, M. G., 67, 145, 148, 149, 155
Boothroyd, C. W., 66
Bose, R. G., 32
Boswall, R. L., 96
Brian, P. W., 168, 205
Bristol, B. M., 197
Broadfoot, W. C., 155, 177
Brown, J. C., 41, 56, 181

Buchanan, R. E., 32
Bunt, J. S., 11, 191
Burkholder, W. H., 119
Butler, E. E., 26, 62, 145, 146, 197
Butler, F. C., 125
Buxton, E. W., 133, 134, 135
Caldwell, R., 120
Campbell, W. A., 64, 85
Carter, H. P., 140
Chacko, C. I., 124
Charpentier, M., 32
Chase, F. E., 102
Chesnin, L., 97
Chesters, C. G. C., 17, 24
Chinn, S. H. F., 62, 122, 174
Cholodny, N., 40
Clark, F. E., 48, 49
Conn, H. J., 13, 36, 187, 194
Cook, F. D., 51, 149, 150
Cook, R. J., 137, 183
Cooke, R. C., 31
Cooper, D. J., 85
Corke, C. T., 13, 194
Couch, H. B., 82, 83, 84
Couch, J. N., 27
Crook, P., 13, 194
Crosse, J. E., 148, 149
Curl, E. A., 26, 147, 183, 186, 197

Damirgi, S. M., 9
Davey, C. B., 9, 67
Davies, B. E., 98
Davies, F. R., 74
Deems, R. E., 183

243

Dick, M. W., 27
Dickey, R. S., 199
Dimock, A. W., 85
Dobbs, C. G., 121
Drechsler, C., 76, 145
Duddington, C. L., 30, 199
Dukes, P. D., 87

Eckert, J. W., 76
Edgell, J. W., 10
Eggins, H. O. W., 33, 197
Elkan, G. H., 106
El-Nakeeb, M. A., 13, 193
Eren, J., 31, 43
Erwin, D. C., 76
Evans, E., 116, 117, 165, 168, 184
Evans, G., 71, 72

Farley, J. D., 9
Faust, M. A., 31
Ferguson, J., 145
Ferris, J. M., 85
Flentje, N. T., 130, 136, 137
Flint, L. H., 33
Flowers, R. A., 64, 198
Freeman, T. E., 140, 141

Gabe, D. R., 41
Gamble, S. J., 9
Gams, W., 21
Garrett, S. D., 79, 120, 124, 125
Georgopoulos, S. G., 185
Ghabrial, S. A., 89
Gilbert, W. W., 77
Gilmore, A. E., 3
Gilmour, C. M., 103
Golebiowska, J., 3
Goodman, J. J., 181
Goos, R. D., 29
Gopalkrishnan, K. S., 181
Goth, R. W., 66
Gottlieb, D., 168, 171, 205
Green, R. J., 71, 202
Gregory, K. F., 162, 170, 206

Grossbard, E., 4, 170

Haas, J. H., 63
Hack, J. E., 24
Halvorson, H. O., 32, 33, 126
Hams, A. F., 177
Hansen, H. N., 74, 91, 189
Hanson, A. M., 27
Harley, J. L., 51
Harris, P. J., 37
Harrison, M. D., 71, 85
Harrison, R. W., 80
Harvey, J. V., 27
Healy, M. J. R., 2
Henderson, M. E. K., 33
Hendrix, F. F. Jr., 63, 79, 198
Henis, Y., 182
Herr, L. J., 141
Hessayon, D. G., 167, 173
Hesseltine, C. W., 29
Hildebrand, A. A., 183
Hine, R. B., 67, 102
Hirte, W., 9
Hoagland, D. R., 206, 207
Hofmann, E., 108
Hornby, D., 9
Hsu, S. C., 154
Hubert, E. E., 78

Isaac, I., 155

Jackson, M. L., 96
Jackson, R. M., 121
James, N., 10, 11, 26, 93, 192
Jensen, H. L., 194
Jensen, V., 9
Johansen, D. A., 46
John, R. P., 34
Johnson, H. G., 126
Johnson, L. F., 17, 25, 26, 76, 83, 89, 154, 155, 158, 177, 178, 184, 197
Jones, P. C. T., 25, 38
Joslyn, D. A., 158

Katznelson, H., 49, 135
Kaufman, D. D., 82, 93, 95, 186
Kelner, A., 140
Kerr, A., 59, 138
King, C. J., 62, 182
Kitzke, E. D., 27
Kleczkowska, J., 149
Klemmer, H. W., 30, 66
Klotz, L. J., 65
Koike, H., 141, 173, 174
Kommedahl, T., 180
Kruger, W., 167, 172, 181
Kuhlman, E. G., 60, 79, 200
Kumar, D., 43
Kuster, E., 195

Lawrence, C. H., 12, 77
Leach, L. D., 68
Ledingham, R. J., 43, 62, 180
Legge, B. J., 120
Lehner, A., 40
Lenhard, G., 108
Leo, M. W. M., 96, 97
Levisohn, I., 29
Lingappa, B. T., 13, 123, 194
Lipman, C. B., 3
Liu, S., 180
Lloyd, A. B., 70
Lochhead, A. G., 10, 183, 192, 205
Loo, Y. H., 163
Louw, H. A., 51
Lucas, R. L., 125
Ludwig, R. A., 184
Luttrell, E. S., 20

MacWithey, H. S., 19
Maier, C. R., 182
Makkonen, R., 145, 147
Maloy, O. C., 126
Manka, K., 4, 24
Mankau, R., 177
Mankau, S. K., 145
Martin, J. P., 26, 196, 197

Martinson, C. A., 67, 127, 201
Matturi, S. T., 125
McCain, A. H., 64, 69
McDaniel, H. R., 34
McKee, R. K., 127
McKeen, W. E., 76, 181
McLaren, A. D., 91
Mead, H. W., 73
Menna, M. E. di., 28
Menzies, J. D., 69, 71, 77, 167, 202, 203
Meridith, C. H., 155
Miller, J. J., 28
Miller, P. M., 190
Minderman, G., 44, 45
Mirchink, T. G., 167, 171
Mitchell, R., 173, 180, 182
Mixon, A. C., 120
Montegut, J., 24
Moreau, R., 56
Morrison, R. H., 60
Morton, D. J., 155
Moubasher, A. H., 184
Mueller, K. E., 19, 67
Munnecke, D. C., 91

Nadakavukaren, M. J., 71
Nash, S. M., 2, 37, 60, 200
Nelson, G. A., 43, 119
Nelson, N., 111
Newcombe, M., 125
Nicholas, D. P., 46, 47
Norton, D. C., 116
Novogrudsky, D., 180

Ocana, G., 63
Ohms, R. E., 30
Okafor, N., 125
Old, K. M., 120
Olsen, C. M., 185

Pady, S. M., 25
Paharia, K. D., 14
Papavizas, G. C., 26, 52, 53, 54, 61,

68, 70, 87, 88, 125, 182, 195, 200, 203
Parker, F. W., 92, 99
Parkinson, D., 50, 51, 54, 56
Parmeter, J. R. Jr., 61
Patel, M. K., 74
Patrick, Z. A., 140, 155
Peterson, E. A., 155
Pisano, M. A., 191
Porter, J. N., 12, 13, 193
Porter, L. K., 108
Pratt, R., 158, 161, 162
Pridham, T. G., 155

Rangaswami, G. 167
Ranney, C. D., 85
Raper, K. B., 187, 188
Rautenshyeyn, V. I., 151
Reeve, R. C., 96
Rehacek, Z., 12
Reyes, A. A., 50
Rhoades, H. L., 145, 177
Richards, L. A., 96, 99
Riker, A. J., 187, 188, 189, 190
Risbeth, J., 78
Robertson, N. F., 79, 80
Robinson, J. B., 151, 152
Rodriguez-Kabana, R., 15, 102, 104, 108, 112, 113, 114, 115
Rogers, C. H., 62
Rohde, P. A., 190
Rohde, R. A., 176
Rose, R. E., 2
Rossi, G., 40
Rouatt, J. W., 183
Rovira, A. D., 131, 135, 136, 206
Rozhdestvenskii, V. A., 3
Russell, P., 29, 78, 202

Schenck, N. C., 14
Schmidt, E. L., 43
Schmitthenner, A. F., 2, 26, 66, 196, 201

Schreiber, L. R., 119
Schroth, M. N., 59, 135, 137, 199
Scott, W. W., 99
Sewell, G. W. F., 26
Shapiro, R. E., 85
Sharvelle, E. G., 165
Sheard, R. W., 96
Sherwood, R. T., 75
Shive, J. W., 207
Singh, K. G., 52
Singh, R. S., 65
Skinner, F. A., 36
Skujins, J. J., 108, 109
Slagg, C. M., 179
Sleeth, B., 76
Smith, J. G., 161
Snyder, W. C., 74, 188
Somogyi, M., 111
Soulides, D. A., 168
Stansly, P. G., 142, 143
Starkey, R. L., 55
Staten, G., 185
Steele, A. E., 85
Steiner, G. W., 26
Stessel, G. J., 142, 144, 155, 204
Stevenson, I. L., 9, 155, 170, 174, 177
Stoner, M. F., 61
Stotzky, G., 4, 102, 104
Stover, R. H., 52, 75, 118
Streets, R. B., 62, 63
Strugger, S., 39
Szybalski, W., 164, 165

Tchan, Y. T., 42
Teakle, D. S., 81
Teliz-Ortiz, M., 156
Thies, W. G., 60
Thomas, A. D., 38
Thornton, H. G., 11, 193
Thornton, R. H., 17
Timonin, M. I., 49, 137
Tisdale, S. L., 89

Tolmsoff, W. J., 30
Tribe, H. T., 32, 125, 146
Tsao, P. H., 70, 77, 126, 128, 144, 195, 202
Tuite, J., 38, 126
Tveit, M., 179, 180, 181

Umbreit, W. W., 85, 102
Unger, H., 3

Van Bavel, C. H. M., 87
Vasudeva, R. S., 166, 169
Vujicic, R., 118

Waid, J. S., 41, 52, 80, 121
Waksman, S. A., 11, 15, 140, 144, 158, 161, 162, 192, 193, 204
Warcup, J. H., 15, 20, 24, 28, 29, 73
Warren, H. B. Jr., 204
Watson, R. D., 21, 61
Weber, D. F., 69
Weinberg, E. D., 165

Weinhold, A. R., 183
Weindling, R., 145, 185
Wilhelm, S., 77, 129
Williams, G. H., 121
Williams, L. E., 24, 156, 157, 183
Williams, S. T., 13, 22, 23, 195
Willoughby, L. G., 27
Willson, D., 34
Witkamp, M., 172
Wood, F. A., 18
Wood, R. K. S., 155
Worf, G. L., 119
Wright, E., 40
Wright, J. M., 169, 179, 180

Yamaguchi, M., 85, 86
Yarwood, C. E., 70, 77, 81

Zak, B., 80
Zan, K., 65
Zentmyer, G. A., 65, 176
Zvyagintsev, D. G., 39, 40